What Deters and Why

Exploring Requirements for Effective Deterrence of Interstate Aggression

Michael J. Mazarr, Arthur Chan, Alyssa Demus, Bryan Frederick, Alireza Nader, Stephanie Pezard, Julia A. Thompson, Elina Treyger

Prepared for the United States Army
Approved for public release; distribution unlimited

For more information on this publication, visit www.rand.org/t/RR2451

Library of Congress Cataloging-in-Publication Data is available for this publication.
ISBN: 978-1-9774-0064-2

Published by the RAND Corporation, Santa Monica, Calif.
© Copyright 2018 RAND Corporation
RAND® is a registered trademark.

Cover: Three ROK soldiers watching the border at Panmunjeom in the DMZ between North and South Korea/Henrik Ishihara via Wikimedia Commons (CC BY-SA 3.0)

Limited Print and Electronic Distribution Rights

This document and trademark(s) contained herein are protected by law. This representation of RAND intellectual property is provided for noncommercial use only. Unauthorized posting of this publication online is prohibited. Permission is given to duplicate this document for personal use only, as long as it is unaltered and complete. Permission is required from RAND to reproduce, or reuse in another form, any of its research documents for commercial use. For information on reprint and linking permissions, please visit www.rand.org/pubs/permissions.

The RAND Corporation is a research organization that develops solutions to public policy challenges to help make communities throughout the world safer and more secure, healthier and more prosperous. RAND is nonprofit, nonpartisan, and committed to the public interest.

RAND's publications do not necessarily reflect the opinions of its research clients and sponsors.

Support RAND
Make a tax-deductible charitable contribution at
www.rand.org/giving/contribute

www.rand.org

Preface

This report documents research and analysis conducted as part of a project entitled *What Deters and Why: Lessons of Deterrence Theory and Practice for U.S. Army Forces and Capabilities*, sponsored by the Office of the Deputy Chief of Staff, G-3/5/7, U.S. Army. The purpose of the project was to provide the U.S. Army with an analysis of fundamental deterrence theory and its potential applications to challenges of interstate aggression, including a specific case study of deterring Russian aggression in Europe.

The Project Unique Identification Code (PUIC) for the project that produced this document is HQD177483.

This research was conducted within RAND Arroyo Center's Strategy, Doctrine, and Resources Program. RAND Arroyo Center, part of the RAND Corporation, is a federally funded research and development center (FFRDC) sponsored by the U.S. Army.

RAND operates under a "Federal-Wide Assurance" (FWA00003425) and complies with the *Code of Federal Regulations for the Protection of Human Subjects Under United States Law* (45 CFR 46), also known as the Common Rule, as well as with the implementation guidance set forth in DoD Instruction 3216.02. As applicable, this compliance includes reviews and approvals by RAND's Institutional Review Board (the Human Subjects Protection Committee) and by the U.S. Army. The views of sources utilized in this study are solely their own and do not represent the official policy or position of DoD or the U.S. Government.

Contents

Preface iii
Figures vii
Tables ix
Summary xi
Acknowledgments xvii
Abbreviations xix

CHAPTER ONE
Understanding Deterrence and Dissuasion 1
Defining Deterrence and Dissuasion 2
A Comprehensive Approach to Preventing Aggression: Strategies of Dissuasion 6
Key Distinctions 7
Intentions, Rationality, and Perception 10
The Choice to Initiate Aggression: Messy, Gradual, and Emergent 13

CHAPTER TWO
Effective Deterrence and Dissuasion: A Framework for Analysis 15
Challenges in Determining What Prevents Aggression 16
Category One: How Intensely Motivated Is the Aggressor? 17
Category Two: Is the Defender Clear and Explicit Regarding What It Seeks to
  Prevent and What Actions It Will Take in Response? 22
Category Three: Does the Potential Aggressor View the Defender's Threats as
  Credible and Intimidating? 23
Summary and Proposed Framework for Deterrence Effectiveness 34

CHAPTER THREE
**Evaluating and Revising the Framework: Quantitative and Case Study
  Assessments** 37
Quantitative Assessment of U.S. Extended Deterrence Cases Since 1945 37
Qualitative Assessment: In-Depth Extended Deterrence Case Studies 39
A Revised Framework of Factors Associated with Deterrence Success 53

CHAPTER FOUR
Applying the Revised Framework: Deterring Russia in the Baltic Region 55
How Motivated Is Russia? 57
How Clear and Explicit Is the U.S. Deterrent Message? 63
Is the U.S. Deterrent Message Credible and Convincing? 70
Conclusion 79

CHAPTER FIVE
Conclusions, Recommendations, and Implications for the U.S. Army 87
Recommendations 89
Implications for the U.S. Army 92

APPENDIXES
A. Quantitative Analysis: Cases of U.S. Extended Deterrence Since 1945 93
B. Qualitative Case Study Analyses: Berlin 109
C. Qualitative Case Study Analyses: Deterring Saddam, 1990 131
D. Qualitative Case Study Analyses: NATO's Northern Flank in the Cold War 145
E. Qualitative Case Study Analyses: Russian Aggression Against Georgia 165

Bibliography 189

Figures

1.1. Deterrence and Dissuasion 7
A.1. Regional Distribution of Deterrence Success in the Face of a Highly Motivated Aggressor 103
C.1. Timeline of Events in 1990 Leading Up to the Iraqi Invasion of Kuwait 133
D.1. The GIUK Gap and the Strategic Situation of the Northern Flank 147
D.2. U.S. Military Personnel Permanently Stationed in Northern Europe, 1950–1992 148
D.3. The Position of Kolsås and Murmansk Relative to Finnmark 150
E.1. Map of Georgia 166

Tables

2.1.	Literature Review: Initial Set of Key Variables	35
3.1.	Key Variables	54
4.1.	Application of the Deterrence Framework to Russia in the Baltic States	80
4.2.	Recommendations for U.S. Deterrent Posture in the Baltics	84
5.1.	Conclusions: Key Variables Governing the Success of Deterrence	88
A.1.	All Cases	95
A.2.	Cases of Failed Deterrence	96
A.3.	Cases of Successful Deterrence	97
A.4.	Deterrence Success in the Face of a Highly Motivated Aggressor	100
A.5.	Deterrence Success with Limited or Ambiguous Clarity	104
A.6.	Deterrence Success Where There Was U.S. Disadvantage in the Local Balance of Forces	106
A.7.	Quantitative Analysis: Confirmed Variables	107
C.1.	Application of the Framework to the Iraq Case	144
D.1.	Application of the Framework to the Case of the United States Deterring the USSR	162
E.1.	Application of the Framework to the Case of the United States Deterring Georgia	177
E.2.	Application of the Framework to the Case of the United States Deterring Russia	185

Summary

The challenge of deterring territorial aggression, which for several decades has been an afterthought in U.S. strategy toward most regions of the world, is taking on renewed importance. An increasingly belligerent Russia is threatening Eastern Europe and the Baltic States with possible aggression, conventional and otherwise. China is pursuing its territorial ambitions in the East and South China Seas with greater force, including the construction of artificial islands and occasional bouts of outright physical intimidation. North Korea remains a persistent threat to the Republic of Korea (ROK), including the possibility of large-scale aggression using its rapidly advancing nuclear arsenal.

Yet the discussion of deterrence—as a theory and practical policy requirement—has lagged in U.S. military and strategy circles. This study aims to provide a fresh look at the subject in this context, with two primary purposes: to review established concepts about deterrence, and to provide a framework for evaluating the strength of deterrent relationships. For greater focus, the study focuses on a specific category or form of deterrence: extended deterrence of interstate aggression. The study considers the requirements for the United States to deter potential aggressors abroad from attacking U.S. allies or other countries in large-scale conventional conflicts. Examples would include Russian attacks on the Baltic States and a North Korean assault on the ROK. It therefore does not consider requirements for deterring other forms of aggression, such as gray-zone campaigns.

The study stems from a specific research question: What are the requirements of effective extended deterrence of large-scale military aggression? The focus is therefore on the criteria that tend to distinguish successful from unsuccessful efforts to deter interstate aggression. This study answers this question, and generates its framework for the analysis of specific deterrent situations, through several specific analytical components.

First, we conducted an in-depth review of available literature on deterrence in general and the requirements of effective extended deterrence in particular. The resulting discussion, presented in Chapter One, offers a refresher course on the nature of deterrence and its requirements.

From this literature we derived 16 factors that tend to determine whether extended deterrence policies succeed or fail. This framework of criteria, presented in Chapter Two, constitutes our tentative assessment of the criteria for effective extended deterrence.

We then sought to confirm the accuracy and utility of that framework through two lines of analysis. The first was a quantitative analysis of 39 cases of U.S.-led extended deterrence since 1945. We chose five leading criteria from the framework, beginning with its three primary categories, and coded each case against these variables based on historical research. We then applied the results to the framework to evaluate factors that the results seemed to support. This analysis is summarized in Chapter Three; Appendix A offers a detailed discussion of the methodology and findings.

The second line of research was qualitative and historical: We conducted four in-depth case studies of key cases of failed or successful extended deterrence involving the United States since 1945. Chapter Three also summarizes each of these cases, and the full, detailed case studies appear in Appendixes B–E. They include an assessment of efforts to deter various levels of interstate aggression in the Nordic countries during the Cold War; West Berlin during the early Cold War; Iraq and Kuwait in 1990–1991; and the Republic of Georgia in 2008. We were looking primarily for cases that mirrored a significant U.S. deterrence challenge today—extended deterrence in far-flung locations where the local balance of capabilities would be significantly tilted against the United States and its allies. We did not, therefore, consider a case with a large imbalance of power in favor of the United States, though such a favorable imbalance *can* certainly contribute to deterrence in some circumstances. We derived lessons from each case and applied the analysis to the framework to isolate factors upheld and called into question by each case study.

Beyond evaluating that framework, our research highlighted several specific themes about successful extended deterrence, including:

- Potential aggressors' motivations are highly complex and typically respond to many variables whose interaction is difficult to anticipate.
- Generally, opportunism in aggression seems less common than desperation caused by real or perceived threats to security or status.
- Clarity and consistency of deterrent messaging is essential. Half-hearted commitments to allies risk being misperceived.
- A "firm but flexible" approach strengthens, rather than weakens, deterrence; leaving an adversary no way out is not an effective way to sustain deterrence. Compromise and concession are typically part of any version of successful extended deterrence of large-scale aggression.
- Multilateral deterrence contexts are especially dangerous. Deterring an aggressive major power while restraining an ally from taking provocative actions at the same time is extremely difficult.

In sum, this analysis suggests that *aggressor motivations serve as the first, and in some ways decisive, variable for interstate deterrence outcomes.* Weakly motivated aggressors are easy to deter; intensely motivated ones, whose level of threat perception verges on paranoia, can be impossible to deter. This finding supports the broader definition of deterrence we suggest in Chapter One: What the United States seeks to do in these cases, in the wider sense, is to *dissuade* a potential aggressor from violent action. Dissuasion includes threats of what the United States (and others) will do in response. But it also includes policies that reassure a potential aggressor and make an attack unnecessary, and to shape the overarching geopolitical context to make aggression both unnecessary and counterproductive.

This analysis suggests that *clarity in what is to be deterred, and how the United States will respond if deterrence fails* is the second essential element of a successful deterrent posture. Lack of clarity invites opportunistic aggression and provides fuel for wishful thinking for highly motivated aggressors; and there are no identifiable cases of failed extended deterrence in which the United States was entirely clear in its interests and intent. In most cases, the United States has fortified that clarity through concrete actions such as alliances, repeated senior-level reiterations of U.S. promises, military exercises and training programs, and deployment of at least symbolic military forces. Clarity backed up by concrete evidence of commitment, including substantial military capabilities, is the foundation stone of effective extended deterrence.

Finally, we applied these findings, and in particular the key deterrence variables, to the current situation in the Baltics to assess the health of the U.S. and North Atlantic Treaty Organization (NATO) policy of deterring Russian aggression against those states. Broadly speaking, in terms of Russia's degree of motivation to undertake aggression and the level of U.S. clarity and commitment, as well as many other of the variables from the set of the leading 12, we find that U.S. and NATO deterrence policy is currently healthy and meets many of the criteria for successful deterrence.

The criteria do, however, suggest areas where it could be improved, including both steps to reinforce the clarity of U.S. messaging and to enhance the capabilities of U.S. and allied military forces. In that analysis, we stress a possible distinction in the broad strategy of enhancing deterrence—to understand whether U.S. and NATO efforts are aiming at achieving objective local military superiority or at deeply complicating Russian planning for any attack. These two distinct objectives may recommend slightly different programs of force enhancement.

Recommendations

In its focused application, this analysis has dealt primarily with one case study—the Baltic region. However, it also carries general implications for the practice of deterrence more broadly. These include a number of fundamental principles of extended deterrence.

Principle One: The United States should carefully assess the national interests and motives of any potential aggressor and seek to ease security concerns that could lead to aggression. This recommendation is essentially to implement the "firm but flexible" approach to deterrence, working to reduce urgent security fears that could lead to aggression through desperation. It does not imply self-destructive concessions or any form of appeasement, but instead simply acknowledges one fundamental truth about deterrence and dissuasion: It is possible for potential aggressors to be provoked into attacking. Deterrence fails as often because of aggressor paranoia and desperation as it does because of opportunistic adventurism.

Principle Two: The United States should work diplomatically to create a geopolitical context hostile to aggression. The surrounding context and the range of political and economic consequences of aggression can help to shape a potential aggressor's motives. In all such cases, the United States can use regional diplomacy to bolster the credibility of a commitment to respond to aggression.

Principle Three: The United States should seek clarity in the actions it pledges to deter and the general scope of its promised reaction to aggression. From the standpoint of U.S. policy, extended deterrence rests most significantly on a foundation of clarity of intent. Where possible, the United States should be specific and clear about actions it is committed to deter and likely actions it would take if that commitment is challenged. In the Baltic case, for example, the implications are relatively straightforward: This principle recommends continuous and unambiguous signaling from the U.S. president and senior officials that these countries are covered by Article V of the North Atlantic Treaty and that the United States intends to stand by its commitment to that treaty.

Principle Four: The United States should take specific steps that reinforce key criteria for successful deterrence as outlined in this study's framework. This study has confirmed 12 variables as being closely associated with the success or failure of extended deterrence. In each case, the study has identified specific action that can reinforce the deterrence-supporting elements of each of those variables. The study provides a detailed menu of options to bolster deterrence in specific cases.

Principle Five: The United States should deploy or support sufficient local capabilities to signal the seriousness of its commitment, to deprive an aggressor of a possible fait accompli, and to offer enough defensive power to assure that the conflict will not remain limited—without employing specific capabilities or postures that the other side will view as immediately threatening to its security. This study concludes that a demonstrated ability to win the local fight is not *necessary* for deterrence, but a strong enough capability to forestall a fait accompli is still a useful, and often critical, component of deterrence. This again points to a possible distinction in European deterrence and reassurance initiatives—between a process designed to create war-winning local capabilities and one aimed at creating enough complications to deprive Russia of any vision of a short or isolated campaign. The latter approach would seek to strengthen local forces and also develop multiple means of projecting power into the Baltics on short notice, in part to convey the certainty that the war would spread very quickly and impose massive costs.

This portrait of the factors governing success and failure in extended deterrence of interstate aggression carries significant implications for the U.S. Army. These include the following:

- Ground forces have a critical role to play in sustaining deterrence in general, and in Eastern Europe in particular. They serve as a powerful signal of U.S. commitment and resolve, and they underline the clarity of U.S. threats.
- However, effective deterrence demands a comprehensive integration of instruments of national power, beginning with diplomacy and negotiations.
- The ability of local U.S. forces to win a contest outright is of less importance than the presence of some forces, wider steps to bolster deterrence, and a minimum ability to forestall defeat to assure that the conflict will remain contained. Nonetheless, additional U.S. capabilities in the region, with some deployed in the Baltic States, would help rule out the potential for any rapid and limited strike by Russian forces.
- Special Operations Forces (SOF) and train-and-advise forces can play an important role in enhancing the defensive capabilities of partner nation forces. Investments in the defensive capabilities of allies and partners—to include support to whole-of-government comprehensive defense plans and societal resilience and resistance efforts—can be more cost-effective than deploying U.S. forces.
- Ground forces provide multiple options to enhance deterrence while sidestepping the provocative, deterrence-threatening risks of some other systems, such as strategic strike capabilities.

Acknowledgments

The research team would like to thank its sponsors, especially MG William Hix and Mr. Tony Vanderbeek, for their support throughout the course of the project. We would also like to express our appreciation to Tim Bonds, director of RAND Arroyo Center, and Sally Sleeper, director of Arroyo Center's Strategy, Doctrine, and Resources Program, for their support of our work. We have benefited from the comments and insight of two RAND colleagues, Samuel Charap and Karl Mueller. We would also like to thank Steve Flanagan of RAND, and Patrick Morgan of the University of California–Irvine, who provided extensive and useful comments on the study during the peer review process.

Abbreviations

AFNORTH	Allied Forces Northern Europe
CIA	Central Intelligence Agency
CIS	Commonwealth of Independent States
CONUS	Continental United States
DIA	Defense Intelligence Agency
DMZ	demilitarized zone
DPRK	Democratic People's Republic of Korea
EU	European Union
GIUK gap	Greenland-Iceland-UK gap
ICBMs	intercontinental ballistic missiles
MAP	Membership Action Plan
MC 14/3	Military Committee document 14/3
MIDs	militarized interstate disputes
MRDs	motorized rifle divisions
NATO	North Atlantic Treaty Organization
NSC	National Security Council
NSD 26	National Security Directive 26
NWFZ	nuclear-weapons-free zone
OPEC	Organization of the Petroleum Exporting Countries
PRC	People's Republic of China

REFORGER	Return of Forces to Germany
ROK	Republic of Korea
SACEUR	Supreme Allied Commander Europe
SOF	Special Operations Forces
SSBN	nuclear-powered ballistic missile submarine
UAE	United Arab Emirates
WMD	weapons of mass destruction

CHAPTER ONE
Understanding Deterrence and Dissuasion

This study examines strategies for preventing interstate aggression in areas far removed from the deterrer's home territory, a task commonly referred to as extended deterrence of territorial aggression. This chapter reviews existing literature on deterrence to define its character and essential challenges. Chapter Two then builds a framework of the factors that determine deterrence success or failure.[1] This analysis eventually leads to a larger focus: the role of deterrent threats within more comprehensive *strategies of dissuasion* designed to prevent aggression. These two concepts are not equivalent, and we do not treat them as such; deterrence is a component of the larger concept. But our analysis finds that only by analyzing deterrence in this larger context can U.S. decisionmakers gain an accurate appreciation for the circumstances in which various strategies will succeed or fail.[2]

This chapter first seeks to define deterrence as a strategic process. It examines various types and forms of deterrence, and discusses their implications for developing effective deterrence strategies. The chapter emphasizes the importance of perception and psychology and the interactive nature of the deterrence task. In the process, it calls attention to the way in which wartime decisionmaking typically occurs—as emergent, gradual, comparative choices under many constraints.

The analysis points to three critical themes to understand the deterrence challenge and to build effective strategies. First, *the threat-making components of any deterrence strategy are much less effective without attention to the larger geostrategic context.* Ultimately, a deterrer is trying to alter the cost-benefit calculus of conflict for a potential aggressor in broad and encompassing terms. Even the most credible threats can fail if the aggressor sees no way to serve its interests outside war. In order to succeed,

---

[1] It is therefore an analysis of state behavior. There is a rich literature on the applicability of deterrence to nonstate actors, but the present study will not evaluate it.

[2] This formulation creates some difficulties with terminology. Because this is a study of deterrence, and because we argue for a broad understanding of that concept, we believe *deterrence* remains the appropriate term, and we use it as the default throughout this report. Substantively, we have in mind a wider notion that could accurately be described as *dissuasion*. We clarify the distinction between the two later in this chapter.

deterrent policies must usually be nested in a larger strategy designed to make aggression as unnecessary to a potential attacker as it is costly.

Second, *deterrence and dissuasion must be conceived primarily as an effort to shape the thinking of a potential aggressor.* Deterrent policies are often viewed through the deterrer's lens—actions that *it* takes to raise the costs and risks of an attack, and ways it measures likely operational outcomes through its own campaign modeling. But the value of those steps depends entirely on their effect on the perceptions of the state considering aggression. Any strategy to prevent aggression must begin with an assessment of the interests, motives, and imperatives of the potential aggressor, including its theory of deterrence (including what it values and why).

Third, *the initiation of conflict through aggression typically results from a complex decision process that unfolds gradually and is not characterized by a single decision point.* Decisions for war are emergent processes rather than moments in time. To be sure, opportunistic adventurism does occur. But many states undertake aggression because they end up feeling they have no choice, and they get to that point through a circuitous process of thinking that seldom begins with that outcome in mind. The challenge of deterring territorial aggression must be understood in this context: not as a strategy to influence a specific, one-time, rational cost-benefit calculus, but as a means of shaping the perceptions of the potential aggressor through a messy, tortuous process that can unfold over months or years.

Defining Deterrence and Dissuasion

In one sense, the concept of deterrence is deceptively easy to define. To *deter*, in the dictionary sense, means to "discourage or restrain from acting or proceeding." It involves preventing some entity from taking an unwanted action—as distinguished from *compel*, which means "to force or drive, especially to a course of action." Compelling someone is making them do what they might not want to do; deterring them is preventing them from taking some action they might want to take. Deterrence can be practiced by denying a potential aggressor pathways to conduct an attack successfully, through defensive measures that credibly deny a potential aggressor its objectives, or by threatening severe punishments if it does engage in aggression. Or, most often, deterrence is practiced through some combination of these means.[3]

---

[3] A number of definitions specify what it is that a deterring state is trying to prevent. The most common specified objective is to prevent military action or aggression of some form. As André Beaufre, *Deterrence and Strategy*, New York: Praeger, 1965, p. 24, suggests, "The object of deterrence is to *prevent* an enemy power *taking the decision* to use armed force" (emphasis in the original). Paul K. Huth, *Extended Deterrence and the Prevention of War*, New Haven, Conn.: Yale University Press, 1988, p. 15, defines deterrence as "a policy that seeks to persuade an adversary, through the threat of military retaliation, that the costs of using military force to resolve political conflict will outweigh the benefits." The present study focuses on interstate aggression, so for our purposes, that is the objective of the deterrent policies we will consider.

This general idea can be understood in a range of ways. Perhaps the narrowest definition, but also one of the most common, holds that deterrence refers to *military tools of statecraft*.[4] In this conception, *deterrence* refers to the strategy of using the threat of military response to prevent a state from taking an action it feels tempted to take.[5]

A second and broader definition expands the scope of possible deterrent policies beyond military responses to any tools of statecraft. A state can deter via threats of economic sanctions, diplomatic exclusion, or information operations. It can promise to rally the international community to impose a reputational cost. This second understanding of the concept continues to focus on responses to unwanted acts, but includes means beyond military ones.[6]

In this broader understanding, threats can be tied to things such as economic prosperity, or even considerations of status and legitimacy. Some historians suggest that such softer forms of threat have worked in the past. Paul Schroeder has noted that it is not right to say that the nineteenth-century Concert of Europe "offered no effective means of deterrence—quite the contrary. Deterrence under the Vienna system took the form of moral and legal political pressure, the threat that reckless or unlawful behavior would cost the offending state its status and voice within the system, leading to its isolation from it and the attendant loss of systemic rewards and benefits." He argues that this form of deterrence was "highly effective" in many cases.[7]

Both of those first two concepts, however, agree with the basic definition offered by Paul Huth and Bruce Russett that deterrence is "dissuasion by means of threat."[8] It can be based on "the capability of defense denying the adversary its immediate

---

[4] See, for example, Lawrence Freedman, *Deterrence*, London: Polity Press, 2004, pp. 26–27, 36–40.

[5] In one of the classic studies of deterrence, Patrick Morgan focused on these force-based elements of the concept: "Deterrence involves manipulating someone's behavior by threatening him with harm. The behavior of concern to the deterrer is an attack; hence, deterrence involves the threat of force in response as a way of preventing the first use of force by someone else"; Patrick M. Morgan, *Deterrence: A Conceptual Analysis*, 2nd ed., Beverly Hills, Calif.: Sage Publications, Inc., 1983, 11. Morgan specifically argues against broadening the definition to nonmilitary tools in Patrick M. Morgan, *Deterrence Now*, Cambridge: Cambridge University Press, 2003, 119.

[6] Bruce M. Russett, "The Calculus of Deterrence," *Journal of Conflict Resolution*, Vol. 7, No. 2, June 1963, pp. 97–98, points to the "numerous nonmilitary ways" that states can deter. Morgan, 1983, p. 19, offers a somewhat different, more encompassing definition: "Deterrence is the use of threats of harm to prevent someone from doing something you do not want him to do." Morgan makes no reference to aggression as the focus of deterrence, or force as the necessary tool, though his study overall consistently returns to the narrower definition—the use of force to prevent attack; he discusses the problem of defining deterrence on pp. 20–26.

[7] Paul W. Schroeder, *Systems, Stability, and Statecraft: Essays on the International History of Modern Europe*, New York: Palgrave Macmillan, 2004, p. 51.

[8] Paul Huth and Bruce Russett, "Deterrence Failure and Crisis Escalation," *International Studies Quarterly*, Vol. 32, No. 1, March 1988, p. 30.

objectives" or on "the threat of inflicting heavy punishment in a larger struggle."[9] But either way, this foundational concept of deterrence involves affecting the calculus of risk and cost by threatening either or both the potential for success in the action or other interests of the aggressor.[10]

A third and still broader way of approaching deterrence is to understand the idea of seeking to "discourage or restrain" in its largest possible sense, and to include means beyond threats in deterrence strategies.[11] One definition of deterrence is "a form of preventive influence that rests primarily on negative incentives,"[12] but this leaves open the possibility of employing positive incentives as well.[13] As Alexander George and Richard Smoke note, "In its most general form, deterrence is simply the persuasion of one's opponent that the costs and/or risks of a given course of action he might take outweigh its benefits."[14] This concept suggests that deterrent strategies can include steps to discourage or even make unnecessary an action in order to prevent it.[15]

Taking seriously the threat-plus-inducement understanding of deterrence can obscure the boundaries of the concept. It risks conflating the more precise understanding of deterrence with more encompassing ideas, and turning a discussion of deterrence into an examination of U.S. national security strategy writ large. Definitions that focus on the military elements of deterrence are theoretically precise and point toward very specific policy implications. Nonetheless, there are powerful reasons—especially in the context of current geopolitical trends, at a time when the line between military and nonmilitary strategies is becoming blurred—to view the challenge of deterrence in this broader way.

---

[9] As Robert Jervis, "Deterrence and Perception," *International Security*, Vol. 7, No. 3, Winter 1982–1983, p. 4, similarly suggests, "One actor deters another by convincing him that the expected value of a certain action is outweighed by the expected punishment," with the term "punishment" seeming to imply threats.

[10] Morgan, 1983, p. 37.

[11] See Paul K. Huth, "Deterrence and International Conflict: Empirical Findings and Theoretical Debates," *Annual Review of Political Science*, Vol. 2, 1999, pp. 29, 38; and Freedman, 2004, pp. 55–59.

[12] Jeffrey W. Knopf, "Three Items in One: Deterrence as Concept, Research Program, and Political Issue," in T. V. Paul, Patrick M. Morgan, and James J. Wirtz, eds., *Complex Deterrence: Strategy in the Global Age*, Chicago: University of Chicago Press, 2009, p. 37. Knopf (pp. 40–41) explicitly notes that both military and nonmilitary tools can produce deterrent effects.

[13] One analysis of extended deterrence suggests that four "types of influence" may be relevant to the success or failure of the deterrent threats: reassurances directed to allies; accommodation of the potential aggressor's objectives to remove the incentive to attack; restraint (especially of the protégé state) to reduce crisis instability; and deterrent threats; Timothy W. Crawford, "The Endurance of Extended Deterrence," in T. V. Paul, Patrick M. Morgan, and James J. Wirtz, eds., *Complex Deterrence: Strategy in the Global Age*, Chicago: University of Chicago Press, 2009, p. 289.

[14] Alexander L. George and Richard Smoke, *Deterrence in American Foreign Policy: Theory and Practice*, New York: Columbia University Press, 1974, p. 11.

[15] Morgan, 2003, p. 119, believes that the result is too encompassing.

In real-world situations in which the United States seeks to discourage a state from taking an action, it very often does combine threats and inducements. In cases of nonproliferation, for example, the United States seeks to discourage a state from developing a nuclear capability by threatening consequences (mostly nonmilitary) if it does so. At the same time, the United States also offers possible benefits if that state agrees to constrain its nuclear ambitions, and provides assurances that may ease that state's perceived need for such weapons. Such inducements can address the motives that a state may have for taking a given action; they force the United States to take seriously the preferences of other states, and thus arguably contribute to a more comprehensive and ultimately effective approach.[16]

The role of inducements is also important because it implicitly recognizes that national security choices are nearly always relative, not absolute. A state's leadership typically weighs the costs, risks, and benefits of a given course *as opposed to* others, rather than strictly on that course's own merits. It is often the perceived costs and risks of alternatives to aggression that lead states to brush off deterrent threats and attack. When considering an invasion of Afghanistan in 1979, for example, Soviet leaders were motivated primarily by the perceived risks of allowing a proxy state to escape the communist orbit—an obsession that caused them to downplay the risks of an invasion.

Influencing a potential aggressor's behavior, therefore, is seldom merely about reducing the feasibility and raising the costs of aggression. It must seek to change the calculus of cost and risk across a range of possible alternatives, causing the potential aggressor to prefer a different option from the one that we are trying to deter. Deterrence is in this sense almost always a comparative rather than narrow and singular task.[17]

Even the term *threat* can be misleading, implying a narrower set of strategies than states typically use. Deterrence is just as often produced by assurances (such as treaty commitments) and actions (such as military deployments) designed to convince a potential aggressor that their gambit would fail. Stationing troops in an ally's territory to prevent aggression is not, strictly speaking, a "threat"; it is an action that fortifies a promise—to defend the ally—and creates an implicit threat of escalation.

---

[16] Robert Jervis argues that even effective reward-based strategies may "lie outside the scope of deterrence theory, which deals with punishment. But if this is so, then the theory may be ruling out consideration of an important tool of influence. Unless scholars know the conditions under which these tools cannot be used, they will sometimes apply deterrence theory to cases which it cannot explain. And decision makers who are guided by the theory and do not heed the qualification that the use of rewards lies outside its scope will rely too heavily on threats and force"; Robert Jervis, "Deterrence Theory Revisited: Review Article," *World Politics*, Vol. 31, No. 2, January 1979, p. 295. One approach that emphasizes the importance of understanding potential aggressor motivations is the idea of *tailored deterrence*; see, for example, Barry R. Schneider and Patrick D. Ellis, eds., *Tailored Deterrence: Influencing States and Groups of Concern*, Maxwell Air Force Base, Ala.: U.S. Air Force Counterproliferation Center, 2011; and M. Elaine Bunn, *Can Deterrence Be Tailored?* Strategic Forum No. 225, Washington, D.C.: National Defense University, Institute for National Strategic Studies, 2007.

[17] George and Smoke, 1974, pp. 520–521; Jervis, 1982–1983, p. 13.

In some sense, the distinction here is ultimately semantic. There *is* a military component to preventing aggression, and there is clearly a broader geopolitical and diplomatic component. As Figure 1.1 suggests, it is possible to term the military element *deterrence* and the wider strategy something else. For the purposes of this study, we are concerned with the broader effort to deter and dissuade potential states from undertaking interstate aggression.

A Comprehensive Approach to Preventing Aggression: Strategies of Dissuasion

We take this approach in part because our research suggests that, in the key deterrent relationships around the world today, such a broader perspective is essential to inform the question of which strategies are likely to prevent aggression. It must encompass both the intentions and perspectives of potential aggressors, the variables at work in their cost-benefit calculus about war, and the steps likely to make aggression seem both more dangerous and less necessary—especially relative to other options. In order to understand what will give an aggressor pause, therefore, we must think in terms of what this study defines as *strategies of dissuasion*—that is, efforts to reduce the perceived utility of or need for aggression.

We define *dissuasion* in the context of extended deterrence of interstate conflict as *the use of military or nonmilitary deployments, threats, and assurances to avoid the recourse to aggression by the target state*. In this definition, *dissuasion* encompasses a more comprehensive approach than *deterrence* in its narrower sense, which is primarily about threats. In this study we understand the widest definition of *deterrence* and *dissuasion* as synonymous, and will continue to use the term *deterrence* throughout the report.

The study considers both threats and broader policies that ultimately aim at *dissuading* a potential aggressor from a given course of action by manipulating many of the factors influencing their choice. Only such comprehensive attention to the variables that influence aggression will provide U.S. decisionmakers with an accurate view of the causes of aggression and the steps necessary to forestall it. Again, as Figure 1.1 suggests, we do not suggest that deterrence and dissuasion are one and the same thing—only that U.S. strategy should focus on the broader challenge and include an analysis of the narrower requirements for deterrence as traditionally defined.

The role of a broader concept of dissuasion is especially important because of a theme woven throughout this analysis: the ways in which threat-based deterrence strategies can go tragically wrong and provoke the very conflicts they mean to avoid.[18] They can do so in several ways, primarily by provoking an action-reaction cycle of military actions that either leads directly to conflict or creates a situation so tense that minor

---

[18] Jervis, 1982–1983, p. 3; Robert Jervis, "Rational Deterrence: Theory and Evidence," *World Politics*, Vol. 41, No. 2, January 1989, p. 183.

Figure 1.1
Deterrence and Dissuasion

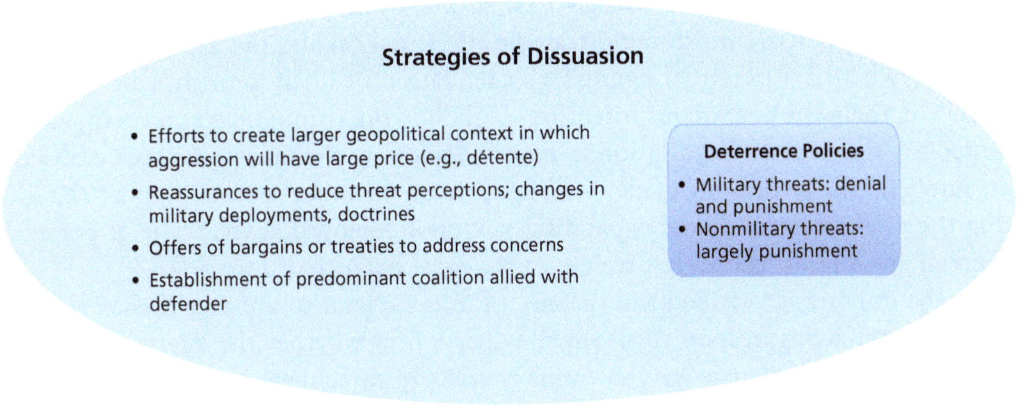

Strategies of Dissuasion

- Efforts to create larger geopolitical context in which aggression will have large price (e.g., détente)
- Reassurances to reduce threat perceptions; changes in military deployments, doctrines
- Offers of bargains or treaties to address concerns
- Establishment of predominant coalition allied with defender

Deterrence Policies
- Military threats: denial and punishment
- Nonmilitary threats: largely punishment

provocations set off major wars. In order to understand what dissuades states from undertaking aggression, we must take into account factors that would cause them to see such aggression as either beneficial or necessary. In some cases, those factors can be deterrent policies themselves.

Key Distinctions

The literature on deterrence makes a number of key distinctions that have implications for understanding deterrence successes and failures.

Deterrence by Denial Versus Deterrence by Punishment

The classic literature distinguishes between two fundamental approaches. *Deterrence by denial* strategies build up military capability to prevent a potential aggressor from succeeding in their attempted attack.[19] Typically this takes the form of building up sufficient local military power to rule out a low-cost fait accompli. At their extreme, these strategies can confront a potential aggressor with the potential for catastrophic loss. In its pure form, deterrence by denial may be indistinguishable from simple defense; a capability to deny is, by definition, a capability to defend, and thus "deterrence and defense are analytically distinct but thoroughly interrelated in practice."[20]

Deterrence by punishment, on the other hand, threatens to impose costs through retaliation that may be unrelated to the aggression itself. Rather than focusing on the

---

[19] Beaufre, 1965, p. 23, argues that in the prenuclear era, a capacity to deter simply meant a capacity to win; he describes the conventional deterrence dynamic as the "dialectic of expectation of victory on the part of the two opponents" (p. 51).

[20] Morgan, 1983, p. 32.

denial of local objectives, it seeks to raise the cost of aggression—even if successful—by threatening other consequences.

Some classic deterrence theorists argue that strategies of denial are inherently more persuasive than punishment strategies.[21] Denial capabilities are more easily recognized, whereas a potential aggressor's calculation of likely punishment "depends largely on their estimate of our intentions."[22] Following through on threats to punish, moreover, risks further escalation, whereas deploying denial capabilities embodies an implied will already in evidence. This line of thinking would support the idea that the most effective denial capabilities are ones deployed at the scene of potential aggression—that is, those that weigh on the local balance of forces.[23]

The importance of the local balance of forces relates to another theme: that the choice to initiate aggression turns on the question of whether the aggressor has—or can convince itself that it has—a military strategy to achieve its particular goals at acceptable cost. John Mearsheimer argues that the effectiveness of conventional deterrence is in part "a function of the specific strategy available to the potential attacker." For example, "if one side has the capability to launch a blitzkrieg, deterrence is likely to fail." On the other hand, deterrence will often work "when a potential attacker is faced with the prospect of employing an attrition strategy, largely because of the associated exorbitant costs and because of the difficulty of accurately predicting ultimate success in a protracted war."[24] Deterrence strategies thus must aim in part at denying a potential aggressor even a wishful-thinking belief that it has a strategy or operational concept that can achieve its goals at low risk and cost.[25]

In practice, few deterrent strategies are either purely denial or purely punishment. States can employ a hybrid approach in which a defender develops some capability for denial but complements it with various threats of additional punishment. Such combined strategies can in theory capture some of the benefits of both denial and punishment, but they complicate the task of assessing the efficacy of a deterrent posture.

Direct Versus Extended Deterrence

A second distinction is more straightforward. *Direct deterrence* consists of efforts to prevent attacks on a country itself—in the U.S. case, on the Continental United

---

[21] See, for example, Huth and Russett, 1988, p. 42.

[22] Glenn H. Snyder, *Deterrence by Denial and Punishment*, Princeton, N.J.: Center of International Studies, 1959, p. 4. As Snyder further notes, "To have an adequate denial capability, preferably one situated near or in a threatened area, is the surest sign we can make to the enemy that the area is valued highly by us" (p. 38).

[23] Snyder, 1959, p. 35.

[24] John J. Mearsheimer, *Conventional Deterrence*, Ithaca, N.Y.: Cornell University Press, 1983, pp. 203–212.

[25] As Beaufre, 1965, p. 53, notes, "The game of conventional deterrence must therefore be played with the enemy's doctrines as a yardstick."

States (CONUS). *Extended deterrence* encompasses discouraging attacks on third parties, such as allies or partners. For example, during the Cold War, direct deterrence involved discouraging Soviet nuclear attack on CONUS; extended deterrence involved preventing Soviet conventional attack on the North Atlantic Treaty Organization (NATO).[26]

Typically, extended deterrence is more demanding than direct deterrence.[27] Few states will doubt that a country will employ every ounce of its strength to defend its own homeland. Yet in many cases, potential aggressors manage to convince themselves that a state will abandon a distant ally or friend in the case of aggression. As Thomas Schelling has noted, there are threats that a defending state would rather not act upon, and weakness in deterrence can emerge when an aggressor believes that the other state will decline such action.[28] Reinforcing extended deterrence then becomes a task of convincing the aggressor that the distant defender will respond automatically.

General Versus Immediate Deterrence

Finally, the theoretical literature distinguishes between *general deterrence*, the day-to-day task of preventing unwanted actions over the long term, and *immediate deterrence*, the more demanding requirement to deter a specific, imminent attack.[29] Under general deterrence, the effects can become internalized or socialized to the point that a potential aggressor ceases actively considering the aggressive action.

General deterrence is held to be easier, largely because the aggressor does not necessarily have a strong motive to take the unwelcome action at any specific moment. During the Cold War, deterring Soviet nuclear aggression against the United States became a form of general deterrence: It pointed to some key military requirements (such as a survivable second-strike capability), but with those roughly in place it was not necessarily seen as a very demanding job outside the context of crises.

Immediate deterrence, on the other hand, can pose quite a challenge because it involves a potential aggressor that has urgent reasons for taking action in the near term—reasons that might cause it to downplay or disregard deterrent threats. Furthermore, once an aggressor is set on a course of action and has begun to take preparatory

---

[26] Huth, 1988, pp. 15–18.

[27] Thomas C. Schelling, *Arms and Influence*, New Haven, Conn.: Yale University Press, 1966, pp. 35–36; Austin Long, *Deterrence, from Cold War to Long War: Lessons from Six Decades of RAND Research*, Santa Monica, Calif.: RAND Corporation, MG-636-OSD/AF, 2008, pp. 13–14; Freedman, 2004, pp. 34–36.

[28] Thomas C. Schelling, *The Strategy of Conflict*, Cambridge, Mass.: Harvard University Press, 1980, p. 123.

[29] Huth and Russett, 1988, p. 30; Freedman, 2004, pp. 40–42; Richard Ned Lebow and Janice Gross Stein, "Deterrence: The Elusive Dependent Variable," *World Politics*, Vol. 42, No. 3, April 1990, pp. 336, 342; Jack S. Levy, "When Do Deterrent Threats Work?" *British Journal of Political Science*, Vol. 18, No. 4, October 1988, pp. 488–489; Huth, 1999, pp. 27–28.

actions (such as deploying the necessary military forces), backing down involves costs and risks that further complicate the deterrent challenge at that point.

Patrick Morgan discusses this distinction in ways that make clear the different requirements generated by immediate versus general deterrence. In a situation of immediate deterrence—which he also refers to as "pure" deterrence—a country has a strong and imminent desire to attack but is held back by direct and credible threats of force. A context of general deterrence is less problematic; potential aggressors are "willing to consider" the use of force if the opportunity should arise, but are in no hurry and have no urgent need to do so. When deterrence works, it is because decisionmakers in the potential aggressor state do not take the option very seriously out of "the expectation that such a policy would result in a corresponding resort to force of some sort by leaders of the opposing state."[30]

This is a much lower bar for deterrence strategies to meet. It merely requires the establishment of a defender's willingness to respond with force, and sufficient cost and risk involved to keep any potential aggressor who is not strongly motivated from acting. Therefore, whether in the Baltic States or elsewhere, it is essential to categorize any specific deterrence challenge as a case either of immediate or general deterrence. The requirements for "what deters and why" will differ significantly depending on which version they reflect.

Our focus in this study blurs somewhat the distinction between general and immediate deterrence. Paul Huth suggests that cases of immediate extended deterrence are more critical because they involve crises and the imminent threat of war; he therefore focuses his study on that specific subcategory.[31] However, long-term U.S. presence in allied countries to deter an ongoing risk of aggression—such as the United States has performed in the Republic of Korea—matches closely the potential requirement in the Baltics, and Huth categorizes such missions as general rather than immediate deterrence. Our own focus includes both subcategories.

Intentions, Rationality, and Perception

Over the last three decades, much of the theoretical literature on deterrence has emphasized that deterrence is an interactive process in which the subjective perceptions of the target state are as important as any objective calculus of deterrent strength.[32] As a theory of cost-benefit calculus, the pure, game-theoretical version of deterrence theory rests on a foundation of the objective, rational evaluation of ends, costs, and risks by

---

[30] Morgan, 1983, pp. 42–44; on the distinction more broadly, see pp. 27–47.
[31] Huth, 1988, p. 18.
[32] Jervis, 1982–1983, p. 4.

a potential aggressor.[33] The rationalist paradigm demands a shared and coherent value system of clearly defined objectives, and a process of weighing the benefits and risks of a course of action in measurable ways.

Yet this assumption is often upset because the success of deterrence lies not in the intrinsic character of the policies but in the degree to which it shapes the view of the target state.[34] The perspectives and cognitive styles of the potential aggressor matter greatly in the degree to which it receives and believes deterrent messages.[35]

History is replete with cases of aggressors' ignoring powerful evidence of risks and costs—including direct deterrent threats by others—to take actions that were apparently self-defeating from the beginning.[36] Japan's decision to go to war in 1941 constitutes a leading case in point.[37] Perhaps the dominant theme of the deterrence literature over the last three decades has been that the value of deterrent messages lies in the eye of the receiver,[38] and that deterrence failures arise in the main from mind-sets and beliefs of a potential aggressor that became immune to deterrence.

Richard Ned Lebow has pointed to the limitations and risks of deterrence as a strategy precisely for such reasons. States considering aggression are often motivated for largely internal reasons, whether related to geostrategic or political calculations. If the reasons are powerful enough, they can render a potential aggressor insensitive to outside influence. "Almost without exception," Lebow argues, crises "could most readily be traced to grace foreign and domestic threats which leaders believed could only be

---

[33] As Schelling, 1980, p. 4, notes, "If we confine our study to the theory of strategy, we seriously restrict ourselves by the assumption of rational behavior—not just of intelligent behavior, but of behavior motivated by a conscious calculation of advantages, a calculation that in turn is based on an explicit and internally consistent value system" (cf. pp. 16–17); he adds that deterrence critically depends on the "rationality and self-discipline on the part of the person to be deterred" (p. 11). See also Morgan, 1983, pp. 79–126; Robert Jervis, Richard Ned Lebow, and Janice Gross Stein, *Psychology and Deterrence*, Baltimore: Johns Hopkins University Press, 1985; Richard Ned Lebow and Janice Gross Stein, "Rational Deterrence Theory: I Think, Therefore I Deter," *World Politics*, Vol. 41, No. 2, January 1989, pp. 208–224; Morgan, 2003, pp. 133–148; and T. V. Paul, "Complex Deterrence: An Introduction," in T. V. Paul, Patrick M. Morgan, and James J. Wirtz, eds., *Complex Deterrence: Strategy in the Global Age*, Chicago: University of Chicago Press, 2009, pp. 5–6.

[34] As Schelling, 1980, p. 160, explains, "A strategic move is one that influences the other person's choice . . . by affecting the other person's expectations on how one's self will behave."

[35] Morgan, 2003, pp. 42–79; Janice Gross Stein, "Rational Deterrence Against 'Irrational' Adversaries?" in T. V. Paul, Patrick M. Morgan, and James J. Wirtz, eds., *Complex Deterrence: Strategy in the Global Age*, Chicago: University of Chicago Press, 2009, pp. 61–70.

[36] For catalogs and descriptions of such factors in a deterrence context, see Jervis, 1982–1983, pp. 19–30; and Jervis, 1989, pp. 198–199.

[37] See, for example, Eri Hotta, *Japan 1941: Countdown to Infamy*, New York: Alfred A. Knopf, 2013.

[38] As Schelling, 1980, p. 3, notes, "To exploit a capacity for hurting and inflicting damage one needs to know what an adversary treasures and what scares him." For an examination of Saddam Hussein's behavior leading to the 1991 Gulf War, see Janice Gross Stein, "Threat-Based Strategies of Conflict Management: Why Did They Fail in the Gulf?" in Stanley A. Renshon, ed., *The Political Psychology of the Gulf War: Leaders, Publics, and the Process of Conflict*, Pittsburgh: University of Pittsburgh Press, 1993, pp. 121–138.

overcome through an aggressive foreign policy."[39] He specifically cites four self-directed motives for aggression: fear of an imminent negative shift in the global context; the need to counteract domestic political instability; the weakness of a specific set of leaders; and competition for power among the elites of a state. He is pessimistic about the ability of deterrence strategies to address any of these motives.[40]

In this regard, the role of individual leaders is often crucial.[41] Different individuals will have different risk tolerance, for example, and distinct reactions to threats. They will have different levels of commitment to aggression. In this line of reasoning, deterrence strategies must therefore aim to influence cost-benefit calculations by specific leaders, not generic "states."

These same perceptual dynamics mean that steps taken to deter can also provoke. Skewed and idiosyncratic perceptions mean that no action has objective meaning—only the meaning that the target of the message reads into it. As a result, the risks of deterrence strategies must be taken as seriously as their potential benefits.

The importance of a potential aggressor's worldview, and the many variables that combine to influence such perspectives, provide a powerful rationale for thinking of the requirements for deterring aggression in broader terms than those of mere threats.[42] A state may consider aggression, or believe it to be an awful necessity, for reasons that are largely immune to influence from classic deterrent threats. In order to understand the basis for aggression and the theoretical requirements for preventing it, therefore, a deterring state must seek to begin with a comprehensive assessment of the aggressor's motives (to whatever degree possible). It can then nest its specific deterrent threats in a wider strategy to influence the cost-benefit calculations of the potential aggressor.[43]

---

[39] Richard Ned Lebow, "The Deterrence Deadlock: Is There a Way Out?" *Political Psychology*, Vol. 4, No. 2, June 1983, p. 334. See also Richard Ned Lebow, "Thucydides and Deterrence," *Security Studies* 16, no. 2, April–June 2007, pp. 163–188.

[40] Huth has partly contested the bounded rationality constraint suggested by Lebow. His findings persuade him that, in the immediate-extended deterrence situations he examined, "short-term military and diplomatic actions do have a strong impact on crisis outcomes," suggesting that it is possible to influence an attacker's objective cost-benefit calculus. Moreover, Huth finds that most states initiated crises or threatening postures for strategic reasons, such as gaining concessions, rather than in service of domestic politics, normative considerations or other indirect motives. See Huth, 1988, pp. 201–202.

[41] Morgan, 1983, pp. 150–151, 153–159.

[42] Jervis, 1989, pp. 292–294, argues that broader strategies to transform aggressor motives are an essential component of any such analysis.

[43] This is a major conclusion in George and Smoke, 1974. "Deterrence should not be viewed as a self-contained strategy," they argue based on the evidence from a number of detailed case studies, "but as an integral part of a broader, multifaceted influence process" (p. 591); and they suggest that theorizing move "from deterrence to inducement" in its fundamental viewpoint (pp. 604–610). They admit that the concept of influence could be synonymous with the study of international relations, but argue—like this analysis—that there is an intervening level of application specifically to conflict or crisis situations. We prefer the term *dissuasion* to describe this broader focus, however, because the goal is stopping an actor from taking an action, whereas the concept of inducement implies a much broader sense of coercion or persuasion.

This study aims to assess the health of deterrent relationships and to provide a framework for evaluating these considerations especially in terms of the intentions of the potential aggressor. Key questions include: How strongly motivated are they to acquire control over the disputed territory? What alternatives do they see to attacking? How do they perceive the general international context, and what implications does that have for their motivation in the specific case? As we will argue, the intentions and degree of motivation of the challenger is the starting point for any analysis of deterrence, and often the decisive factor in determining whether deterrence strategies work or fail. Some aggressors can be so powerfully motivated, for example, that they ignore or downplay even the strongest deterrent signals.

The Choice to Initiate Aggression: Messy, Gradual, and Emergent

Finally, understanding the requirements for deterring aggression requires some attention to the typical character of choices to go to war. The case study literature on deterrence makes clear that these are seldom single-point, objective cost-benefit calculations. While they can be based on clear strategic objectives (rather than being the result of extreme cognitive processes like dangerous wishful thinking), the decision processes tend to have a range of factors that complicate the process of intervening with deterrent threats.

First, they *are usually the outcome of a long-term process of actions and reactions rather than a simple choice.*[44] The decision to undertake aggression is seldom sudden. It is often the final stage of a long—sometimes years-long—process of diplomacy, dialogue, negotiation, threat, and counterthreat. It reflects long-term historical, policy, and personal interactions and clashes that create a unique context for each aggressive action, and it embodies taken-for-granted worldviews that have come to characterize the aggressor state's national security dialogue.

Second, and related, *the decision to undertake aggression is usually the result of an emergent thought process on the part of the aggressor, not a decision made at a definable moment.* In most instances of large-scale territorial aggression, case studies of the decision processes cannot point to a single moment at which "the decision" to go to war was made. More often the conviction that a state must undertake aggression emerges gradually over time, finally becoming an overwhelming sense that action is needed. A common theme of such cases is that participants afterward have a difficult time pointing to the moment when decisionmakers actually made the choice.

---

[44] As Schelling, 1966, p. 98, puts it, one view of deterrence "seems to depend on the clean-cut notion that war results—or is expected to result—only from a deliberate yes-no decision. But if war tends to result from a *process*, a dynamic process in which both sides get more and more deeply involved, more expectant, more and more concerned not to be a slow second in case the war starts," then the challenge of deterrence becomes much more complex.

In such decision processes, signals sent at one moment may have a very different result from ones sent at another. In some cases, a potential aggressor could have already come to a powerful *subconscious* conclusion that territorial aggression is required but not yet made the conscious decision to attack. Deterring that outcome, however, will be much more difficult in such cases than if a decisionmaker is approaching a choice from scratch. In short, in order to be successful, deterrent policies must take seriously the emergent, imperative-driven character of many decisions to undertake aggression.

These insights also endorse a broader rather than narrower conception of what it takes to prevent aggression. The theory of deterrent threats often presumes a specific, discrete form and timing of aggression against which the United States can direct tailored threats. But it can be tougher to deploy such threats against an emergent, more obscure choice. In some cases of ongoing general deterrence, this may be less of a problem; the deterrent threat will remain in place to influence even gradual decisions. But a major challenge is the failure of general deterrence, or the transition from general to immediate deterrence—when an aggressor ceases to be dissuaded by the general threats and moves to more active consideration of aggression.

Such a situation would then call for new, additional, and more urgent deterrent threats. But if the United States never knows when that moment has arrived—indeed, if it never *does* arrive as a discrete period when a clearly identifiable decision has been made—then it will be impossible to calibrate immediate deterrent threats effectively. This suggests, again, that broader approaches to dissuasion that manipulate many variables influencing aggression could be more effective.

CHAPTER TWO

Effective Deterrence and Dissuasion: A Framework for Analysis

The literature on deterrence suggests factors that help to determine success or failure. This chapter summarizes the numerous criteria that emerge from the literature into three major categories that point to the most decisive factors in determining deterrence success: Was the potential aggressor intensely motivated to take action? Was the defender clear and persuasive in communicating what it sought to prevent and the action it would take if an attack occurred? Did the defender convince the aggressor of its capability and willingness to respond?

The relative importance of these three categories will differ from case to case, and perhaps the most important finding of our review of the existing literature is that there can be no simple answer to the question of "what deters and why." That literature does not highlight a few variables that consistently explain extended deterrence success across cases. Instead, a kaleidoscope of influences—beginning with complex aspects of the potential aggressor's mind-set, motivations, and worldview—determine outcomes.

Moreover, what deters in one context will not necessarily deter in another. A key question is *under what circumstances* different strategies for deterrence are most effective, and in what combination. There is simply no right answer to this question that will apply to any given case. Instead, we offer here a framework of key factors that can be used to evaluate the strength or weakness of dissuasion and extended deterrence in specific cases. On the basis of this framework, U.S. decisionmakers could construct a scorecard for current and prospective cases of deterrence. The initial version of this framework appears at the end of the chapter in Table 2.1. Chapter Three will then summarize quantitative and qualitative work—presented in more detail in the appendixes—to evaluate this initial list of variables, and will offer a revised and refined framework based upon that work. Chapter Four then employs the revised framework to evaluate the strength of the current deterrent effort to dissuade Russia from aggression in the Baltic States.

As noted above, the empirical and qualitative literature on deterrence points to three fundamental categories of factors that determine when the United States is effective in deterring aggression. Each is posed as a question:

1. How intensely motivated is the aggressor?
2. Was the United States clear and explicit regarding what it sought to prevent and what actions it would take in response?
3. Did the United States convince a potential attacker of its capability and willingness to respond?

In each of these categories, the literature nominates a number of more discrete variables or criteria that can be used in a framework for analyzing deterrence effectiveness.

Challenges in Determining What Prevents Aggression

A review of the existing literature on deterrence suggests that evaluating "what deters and why" is a tremendously complex challenge. The first two waves of deterrence theory, for example, made confident pronouncements about what would and would not deter an aggressor—without any clear foundation. They were highly inductive and theoretical arguments, and their claims "soon became conventional wisdom even though there was little evidence for the validity of the propositions."[1]

That changed in the 1980s and 1990s, when scholars such as Paul Huth, Robert Jervis, Richard Ned Lebow, Patrick Morgan, and Janice Gross Stein undertook more empirical and deeper case study work on deterrence. However, this work only highlighted the challenge of finding general patterns. So many factors are typically at work in any one deterrence case that it may be impossible to distinguish those factors that led to failure or success. An aggressor might never have intended to attack in the first place. Even if it did, it might have hesitated for reasons unknown to the defender, and unrelated to the substance of the defender's threats.[2] Often we cannot know even in the aftermath of a choice what it was that kept a state from attacking. This is especially true in conditions of general deterrence, where there is no specific moment at which a choice to back down is made; a range of military, political, and economic factors may discourage a potential aggressor from ever coming close to launching an attack, and through a complex and largely mysterious process.[3]

---

[1] Jervis, 1989, p. 289; see also pp. 301–303. He adds, "Perhaps the most startling fact about the development of the theory is the lack of search for supporting evidence" (p. 301). See also George and Smoke, 1974, pp. 2–3, 66–71, 94–95.

[2] Morgan, 2003, p. 122, has argued that "detecting success or failure depends on whether the *threat persuades*. This is very messy."

[3] Jervis, 1989, describes the difficulties in making objective judgments about subjective criteria for cost-benefit outcomes (pp. 187–189) and discusses the difficulties of case selection for deterrence (pp. 193–199).

From a methodological standpoint, moreover, building a truly reliable sample of extended deterrence cases can be difficult.[4] Individual cases betray huge differences, making cross-comparison difficult. Some observers argued that large-N, regression-based studies of deterrence end up making wasted efforts to distinguish signal from noise in contexts where the interactions among variables are often nonlinear and subject to the unique conditions of specific cases.[5]

Even when deterrence succeeds, it can be difficult if not impossible to single out the factors that caused that outcome. In any given scenario, the United States may deploy a dozen or more complementary measures to achieve its goal, some of them threat-based and some fitting into a broader geopolitical strategy. The aggressor may hesitate, responding to U.S. measures in combinations and ways that even the aggressor cannot fully describe. This dynamic may make it impossible to determine just what prevented conflict.

Even the definition of success and failure can be ambiguous.[6] In one sense we can measure the effectiveness of deterrence of interstate aggression via an obvious, binary, objective standard: Did the potential aggressor attack, or not? Yet there is a difference between succeeding in general deterrence and immediate deterrence; a deterring state or coalition can fail at the former (sometimes without even realizing it), leaving a potential aggressor constantly looking for the right moment to undertake an attack that it believes would enhance its position. But the defender can still succeed in the immediate deterrence task of never providing a perceived opportunity to launch the attack.

Available evidence thus emphasizes the contingent, context-dependent nature of deterrence.[7] What deters at one time or against one adversary may not work—indeed, may be counterproductive—at other times or with other states. Our review of key conditions is designed to inform a framework or scorecard for appreciating the general conditions for deterrence success. But which ones are most in evidence or important in a specific case can only be determined by filtering these general factors through a context- and adversary-specific analysis. The best we can do, therefore, is to outline factors that appear, in both case-based and quantitative terms, to be regularly associated with success or failure in efforts to prevent aggression.

Category One: How Intensely Motivated Is the Aggressor?

The intentions of the target state are the beginning point of analysis for any deterrence strategy. An obvious example is a state's degree of aggressive intent: A deterrence

---

[4] Morgan, 2003, pp. 123–129, 152–162.

[5] Francis J. Gavin, "What We Talk About When We Talk About Nuclear Weapons: A Review Essay," *H-Diplo/ISSF Forum*, No. 2, 2014, pp. 11–36.

[6] George and Smoke, 1974, pp. 514, 516–517.

[7] This is a major theme in George and Smoke, 1974; see, for example, pp. 3, 54.

situation does not exist unless the state has interests that could be served by aggression and is seriously considering such a course of action.[8] This seems obvious, but the nature of aggressor intentions is often taken for granted in deterrence analysis. The possession of a *capability* to undertake an attack, as well as conceivable reasons for doing so, is often deemed equivalent to an *intention* to do so, and thus sufficient to create a deterrence situation.[9]

Yet capability to attack, the historical record suggests, tells us little. Motives and intentions are critical. If a state sees little reason to undertake aggression, it will not be hard to deter it; if it has acquired an urgent sense that only an attack will safeguard its interests (as with Soviet views of Afghanistan in late 1979, or the administration of President George W. Bush regarding Iraq in 2001–2002), it may become almost impossible to stop. Reviewing a range of theoretical and empirical literature, in fact, Patrick Morgan concludes that "challenger motivation is the most important factor in deterrence success or failure."[10]

In the broadest terms, a potential aggressor's degree of generalized dissatisfaction with the status quo sets the context for these motives. If the aggressor's dissatisfaction is low, it will be relatively easy to deter on certain territorial issues. If, on the other hand, a state has come to believe that its long-term prospects are not good, it may decide that the risk of war is the only option it has, and go to war regardless of deterrent threats.[11]

Again, these decisions are typically comparative rather than binary. Decisionmakers seldom weigh the cost-benefit calculus of starting aggression in the abstract; they consider the *relative* merits of several alternative courses. If leaders view attacking as less risky or costly than any of the alternatives, they will not be deterred.[12] But this comparative decisionmaking process also suggests, as Thomas Schelling points out,

---

[8] Morgan, 1983, p. 35. As Morgan, 2003, p. 121, notes, "If State B doesn't perceive an attack coming or does not threaten to prevent it, it is not practicing deterrence and we can't learn much from that case. If State A has no intention of attacking the case cannot tell us whether deterrence works or how."

[9] Morgan, 1983, p. 36, explains, "The fact that states may have a constant capability to attack may mislead us into thinking that they must be regarded as constantly about to attack." A related point is that deterrence is less relevant in a world of status-quo powers; see Knopf, 2009, p. 43.

[10] Morgan, 2003, p. 164. See also George and Smoke, 1974, p. 532.

[11] Michael E. Brown, *Deterrence Failures and Deterrence Strategies: Or, Did You Ever Have One of Those Days When No Deterrent Seemed Adequate?* Santa Monica, Calif.: RAND Corporation, P-5842, 1977, p. 2. Morgan, 2003, p. 163, argues that the sum of evidence on deterrence suggests that "[d]eterring a highly dissatisfied challenger is very difficult." See also Janice Gross Stein and David A. Welch, "Rational and Psychological Approaches to the Study of International Conflict: Comparative Strengths and Weaknesses," in Nehemia Geva and Alex Mintz, eds., *Decisionmaking on War and Peace*, Boulder, Colo.: Lynne Rienner, 1997, pp. 51–80, which argues that deterrence works best when a potential aggressor is not in the grip of severe misperception and believes that it has the freedom to exercise restraint. Challengers that feel more aggrieved and desperate are much more difficult to deter.

[12] Mearsheimer, 1983, pp. 62–65; Jervis, 1982–1983, pp. 13–14, 67–68; Jervis, 1989, pp. 197, 202; Huth, 1999, pp. 40–41.

that "the pain and suffering" embodied in the deterrent threats "have to appear *contingent* on their behavior."[13] If it comes to be perceived as a general policy of hostility, threats may lose their ability to be applied to deter specific actions.

Lebow therefore suggests that deterrence strategies might only be reliable in a narrow set of circumstances. He nominates key conditions for strategic contexts in which deterrence can work: when the aggressive state to be deterred is motivated by opportunistic gain rather than fears of strategic or political loss; when its political and strategic situations provide it with the "freedom to exercise restraint"; when it is "not misled by grossly distorted assessments of the political-military situation"; when it is "vulnerable to the kinds of threats" that the deterring state can make; and when deterrent threats can be made before the potential adventurist locks itself into an irreversible conscious or subconscious commitment to aggression. "Unless these conditions are met," Lebow concludes, "deterrence will be ineffective or counterproductive."[14]

Specific measures of an aggressor's motivation can range from strategic variables to internal political considerations. They include the following:

- How dissatisfied is the aggressor with the status quo in general terms?
- Is the aggressor concerned that the strategic situation is turning decisively against it or could do so because of emerging trends?
- What degree of perceived interests does the aggressor have in the disputed territory?
- Does the aggressor have a perceived strategic or political reason to act that is urgent or even desperate?
- In its general foreign policy and strategic culture, does the aggressor display an extreme form of risk-accepting aggressive opportunism?
- Do the aggressor's leaders perceive strong personal, political, or ideological reasons to act?

These are not the only factors that will determine a potential aggressor's motivations, but the existing literature suggests that they are commonly among the most important.

A potential aggressor's motives provide the working materials for our broader conception of deterrence—strategies of dissuasion. Motives point the deterring state to what the aggressor values, why it sees a reason to consider aggression, and what alternatives might satisfy its interests. By assessing these motives and taking actions to address the aggressor's interests, the deterring state can, concurrently, reduce the perceived need for and value of aggression while it raises the costs of such aggression with

---

[13] Schelling, 1966, p. 4.

[14] Richard Ned Lebow, "Deterrence Failure Revisited," *International Security*, Vol. 12, No. 1, Summer 1987, p. 212.

deterrent policies. As noted below, there is empirical support for the proposition that such mixed approaches have the greatest chance of success.

The Uncertainty of Intentions

Some scholars assert that it is impossible to "know" the intentions of an adversary. Even if they can be correctly assessed at one moment, moreover, those intentions can change. Therefore, some contend that capabilities are really all that matter, and that efforts to understand the intentions of a potential aggressor are futile.

Yet there is much that a deterring state can know, even if its awareness is never perfect. It can appreciate the national interests perceived by the aggressor and how those might be at stake in a given territorial claim. It can often have a reasonably good sense of the general strategic perspective of such a state. Does it feel confident or vulnerable? Does it perceive the strategic balance tilting against it, or running in its favor? Through both public documents and secret intelligence, a defender can often have a reasonably clear understanding of the objectives sought by a potential aggressor in a specific situation and how it perceives those objectives to be served by various alternative courses of action.

Indeed, history suggests that a lack of available information about a potential aggressor's intentions is seldom the problem. Before Japan attacked Pearl Harbor in 1941, or China intervened in Korea in 1950, or the Soviet Union invaded Afghanistan in 1979, accurate reporting was available in both open and clandestine channels on many of these subjects. Deterrence and dissuasion failed in part because the United States misperceived the signals, or refused to acknowledge their importance, or decided that the actions necessary to answer the aggressor's concerns were unacceptable. But the literature on deterrence does not support a simple assertion that it is impossible to have a good awareness of an aggressor's intentions.

Reciprocal and Flexible Bargaining Strategies

The importance of aggressor motivations—and the fact that those motives serve larger national security objectives—point to the importance of going beyond threats in deterrence and dissuasion strategies. Typically, even potential aggressors are trying, in their own minds, to enhance their security. Therefore, deterrent threats can increase rather than decrease their perceived motive for attacking, and thus prove counterproductive.

One clear finding in both empirical and case-based research is that hard-edged deterrent threats can be counterproductive on their own. Nuanced combinations of threats and concessions appear to be most associated with deterrence success. As Morgan notes, "Mixing deterrence and conciliation is best—be tough but not bullying, rigid, or unsympathetic."[15] Such strategies can address the potential aggressor's

---

[15] Morgan, 2003, p. 162. Morgan adds that "the strength of the challenger's motivation is crucial—weakening it by concessions and conciliation can make chances of success much higher" (p. 163).

security motivations, as well as raising the perceived costs of aggression. This again points to a basic concept of dissuading, rather than merely deterring, aggression.

The empirical record strongly indicates that states that initiate aggression are not merely opportunistic or aggressive, but are often responding to situations they perceive as highly dangerous. Lebow's case study analysis led him to conclude that states that sparked crises were usually responding to what they saw as an "acute and impending danger," such as a significant negative shift in the balance of power.[16] Alexander George and Richard Smoke suggest that an aggressor's "perception that only force or the threat of it can bring about the desired change" has a critical effect on deterrence outcomes.[17]

Such motivations help to account for why deterrent threats can be as dangerous as they are stabilizing. When undertaken rashly, they can confirm the fears that are driving a state to overturn the status quo.[18] Yet efforts to accommodate a potential aggressor often fail, largely because they have become too convinced of the defender's hostile intent to believe any assurances.[19]

Huth's empirical and case study analysis leads him to emphasize the importance of crisis bargaining behavior to the outcome of deterrence cases. The "sensitivity of a potential attacker to military threats and challenges to its reputation," he suggests, complicate the task of deterring without provoking. Abortive deterrence strategies have the potential not merely to fail in their main goal of forestalling an attack, but also to bring one on by making the aggressor more convinced of the need to strike.[20]

The most effective strategies, Huth concludes, therefore involve "firm but flexible" diplomatic stances and tit-for-tat bargaining approaches. Such strategies are based on reciprocity rather than an attempt to impose outcomes. His empirical survey suggests that reciprocal strategies succeed nearly 80 percent of the time, whereas "a record of intransigence by the defender" reduced the success rate by a third.[21]

---

[16] Lebow, 1983, p. 334.

[17] George and Smoke, 1974, p. 531.

[18] Lebow, 1987, p. 211, argues that "deterrence can be malign by intensifying the pressures pushing leaders toward the use of force." Deterrent threats can "underline the dangers of inferiority, exacerbate fears of strategic vulnerability, and encourage leaders to preempt. This is most likely to occur when leaders believe that the military balance will be even less favorable in the future."

[19] Lebow, 1983, pp. 343–344, is not optimistic about the ability of states to reassure others about their benign intentions. Making the effort can either fail, he suggests, or else generate counterproductive reactions. States in the grip of paranoia about others' intentions will see diplomatic advances as mere distractions from the underlying policy of hostility. Lebow suggests that often a dramatic initiative will be required to shock another side out of its assumptions and open the room for taking the reassurance seriously.

[20] Huth, 1988, pp. 9–11.

[21] Huth, 1988, pp. 75–76, 81.

Category Two: Is the Defender Clear and Explicit Regarding What It Seeks to Prevent and What Actions It Will Take in Response?

A second broad criteria for deterrence success is that the defender must be clear about what it is trying to deter, as well as what it proposes to do in the event that the threat is ignored.[22] If a commitment "is ill-defined," Schelling explains, "and ambiguous—if we leave ourselves loopholes through which to exit—our opponent will expect us to be under strong temptation to make a graceful exit (or even a somewhat graceless one) and he may be right."[23]

Korea in 1950 and Iraq in 1990 provide two powerful examples of the dangers of a lack of clarity. In both cases, the United States refused to make clear the outcomes it sought to deter (or would not accept). This failure left two highly motivated aggressors ample room to convince themselves of their ability to launch a fait accompli that would not provoke a decisive U.S. response. In fact, in each case, Washington gave signals that could have been read as having the opposite message—that it simply did not care that much about the disputed territory. In 1950 it was Dean Acheson's infamous speech drawing a security perimeter in Asia that excluded Korea. In 1990 it was the U.S. ambassador's ambiguous statements to Saddam Hussein that the United States had no official view on his territorial disputes with Kuwait. One of the straightest roads to deterrence failure is to be vague about what the United States will not accept.

Yet there can be something of a dilemma regarding the clarity of threats. Some studies argue that such clarity is essential to deterrence, for without it an aggressor can misinterpret the defender's intent. On the other hand, Schelling argues for the value of a "threat that leaves something to chance"—the manipulation of uncertainty to deter more than a state is directly willing to threaten to defend. Bargaining strategies deal with such situations of ambiguity, playing in the gray area between rigid commitment and no concern at all.

Moreover, the United States is seldom willing to provide absolute clarity for all its international commitments. It must strike a constant balance between clarity and flexibility, leaving itself some room for maneuvering. The Syrian "red line" controversy is a good example of a clear commitment that eventually boxed the United States into responses it arguably would have preferred not to make.

Moreover, the target of the deterrent threats must hear and understand them clearly. A key challenge of deterrent threats is to ensure that the target hears the message "through the din and noise" of world politics.[24] The powerful and effective communication of messages is a prerequisite for effective deterrence.

---

[22] George and Smoke, 1974, pp. 561–565.

[23] Schelling, 1966, p. 48.

[24] Schelling, 1980, p. 11; see also pp. 26–28, 47. Elsewhere (1966, p. 38) Schelling writes, "If he cannot hear you, or cannot understand you, or cannot control himself, the threat cannot work."

The historical record does not suggest that even clear and well-communicated commitments offer a guaranteed route to deterrence success. If they have the slightest ambiguity, they can be challenged. Potential aggressors are always probing the seriousness of commitments, trying to determine how credible they are, and a generalized U.S. commitment during the Cold War did not forestall any challenges.[25]

Deterrence and the Problem of Warning

Effective strategic warning can provide a defender a strong opportunity, before an attack occurs, to reinforce the clarity of a deterrent threat. Such warning offers a chance for a defender to communicate with precision what it will not accept and what it will do in response. To take just one example, had the United States made a clearer deterrent threat in the weeks leading up to August 1990—when it began to see indications that Saddam was considering invading Kuwait—it might have been able to deter the aggression.

But there are two problems with these potential opportunities for clarity. One is the problem of effective warning, particularly as it applies to deterrence. Defenders often miss or willfully ignore signals of impending attack for a variety of reasons that have been extensively researched. The implication of this research is that defenders will often fail to see the chance—or requirement—for more pointed deterrent threats as the aggressor's intentions mount. In other words, defenders often miss the transition from general to immediate deterrence—when general deterrent threats have ceased to be effective and an aggressor has moved to considering immediate aggression.

A second problem is that, even at such moments, raising the stakes with additional clarity may, as do deterrent threats as a whole, provoke rather than restrain the aggressor. An aggressor might decide that it must act quickly, before reinforcements arrive. Even if it works in the moment, the expanded deterrent threat may make later challenges more likely by "hardening [the aggressor's] conviction that the defender is unresponsive to the legitimate interests that lie behind his efforts to obtain a change in the situation."[26]

Failure to heed clear warnings of an impending attack can contribute to deterrence failure. We have therefore included a factor on the framework below that captures this important requirement.

Category Three: Does the Potential Aggressor View the Defender's Threats as Credible and Intimidating?

The cornerstone of the classic theory of deterrence is that the defender must have the capability and will to do what it threatens, and that the potential aggressor must

---

[25] George and Smoke, 1974, pp. 525–526. They argue that, during the Cold War, "communist leaders acted to alter the status quo on many occasions *even* in the face of U.S. commitment" (p. 525, emphasis in the original) at such moments as the Berlin crisis and the Chinese intervention in Korea. "An American commitment per se, therefore," they conclude, "is clearly insufficient to prevent failure of deterrence."

[26] George and Smoke, 1974, p. 579; on the warning problem more generally, see pp. 572–580.

appreciate this.[27] That basic requirement in an extended deterrence context is a function of four things: military and nonmilitary capabilities, especially in the local or disputed territory; perceived resolve and willingness to fulfill threats; the national interests involved in the specific dispute, especially relating to economic and political relationships with the threatened state; and reputation as a strategic actor, largely defined by recent encounters with the potential aggressor state.

At the same time, the potential aggressor must perceive that the defender is threatening something serious and damaging enough to raise the costs of an attack beyond what it is willing to pay. Deterrence can still fail if a threat is credible, as long as that threat is to impose a cost the attacker is willing to bear. This is indeed a theme in some recent criticism of U.S. policies toward such states as Iran and Russia: When the United States threatens yet more international sanctions or diplomatic opprobrium, the target states might fully believe it, yet simply not care.

The scale and seriousness of threats work in a reciprocal relationship with the potential attacker's perceived interests. A deterrent threat has only to be powerful enough to overcome the perceived advantage of aggression. If that advantage is minimal, even modest threats can prevent it. If an attacker has vital national interests engaged in a territorial issue—such as North Vietnam's interests in conquering South Vietnam—U.S. deterrent threats would in theory need to be very extreme in order to prevent unwanted aggression.

The Role of Local Force Balances

The capability to deny an aggressor what it might seek is generally held to be the foundation of deterrence. In line with the theme of influencing the potential aggressor's belief system, however, this requirement is less about merely *having* an objective capability to deny objectives but possessing such a capability *as perceived by the potential aggressor*. It is the aggressor's view of capabilities that will determine success or failure.

Recent research suggests that the immediate balance of forces in the contested territory—specifically, the relative balance of forces between a potential aggressor and its allies and the defending state or states—is one of the most important factors leading to deterrence success.[28] Many studies make a distinction between long-term general military balances and short-term local ones; a defender may be clearly stronger overall, but if an aggressor senses the potential for a fait accompli against a weak area, this general superiority may count for little.[29] Empirical studies support the importance of

---

[27] Paul, 2009, p. 2. See also Knopf, 2009, pp. 47–48.

[28] Mearsheimer, 1983, pp. 24, 62; Huth, 1988, p. 74; Huth, 1999, p. 30; Morgan, 2003, p. 162; Long, 2008, p. 9. Huth, 1988, pp. 75–76, concludes that in cases of deterrence failure "either the defender's forces were not in a position to repulse the initial attack, giving the attacker a decisive advantage in the immediate balance of forces, and/or the defender's capacity to mobilize and reinforce local forces did not decisively alter the short-term balance of forces in its favor."

[29] Huth and Russett, 1988, p. 34.

this variable in determining deterrence outcomes.[30] (A combination of these factors is stronger still; defenders perceived to be dominant in overall strength *and* superior militarily at the point of aggression are rarely challenged.)

Yet this variable alone cannot explain the success or failure of many deterrence cases. In some, notable local weakness was never challenged; in others, aggressors disregarded evidence that the defender was superior and attacked anyway.[31] Decisions for war reflect a complex interaction of variables, of which the assessment of mutual strength is only one. States have fought seemingly hopeless wars when they thought the values at stake were worth it; on the other side of the coin, states with dominant power have sometimes refused to deploy it. The essential factor is the motivation and perception of the aggressor. "Wars rarely start because one side believes that it has a military advantage," Lebow contends. "They occur when leaders become convinced that force is necessary to achieve important goals."[32]

And again, there is, as we have stressed, a sensitive balance between deterrence and provocation. As will be discussed below, actions taken to deter, especially if an aggressor perceives itself to be acting out of defensive motives, can convey an impression that the deterrer is in fact preparing to attack. This is particularly true during a crisis, when multiple stability/instability dilemmas can arise during the execution of deterrent policies.[33]

A related issue is the role of nuclear superiority in determining outcomes of deterrence cases—or, more broadly, crisis interactions. Here the evidence is mixed. Matthew Kroenig examined 52 cases, attributing "victory in nuclear crises" (or deterrence success) to nuclear superiority through its effect on the balance of resolve.[34] Yet in their own empirical analysis, Paul Huth and Bruce Russett find no relationship between nuclear weapons and deterrence success.[35] Vipin Narang finds that assured retaliation

---

[30] Huth and Russett, 1988, pp. 38, 42, 43; Curtis S. Signorino and Ahmer Tarar, "A Unified Theory and Test of Extended Immediate Deterrence," *American Journal of Political Science*, Vol. 50, No. 3, July 2006, pp. 594, 598–599; Levy, 1988, p. 510.

[31] Russett, 1963, pp. 102–103. One complication in the relationship between local military strength—denial capabilities—and a broader threat to retaliate is that if denial forces are less than sufficient for defense, their weakness may be as evident as their potential strength. The deterrent value of punishment, on the other hand, while uncertain, is always present and does not depend on local strength; Snyder, 1959, p. 6. George and Smoke, 1974, p. 530, actually list the "defender's military capability" as a "minor condition" affecting deterrence outcomes, which are "less critical" than the leading ones.

[32] Richard Ned Lebow, "Misconceptions in American Strategic Assessment," *Political Science Quarterly*, Vol. 97, No. 2, Summer 1982, p. 197. Lebow makes the argument at greater length in "Windows of Opportunity: Do States Jump Through Them?" *International Security*, Vol. 9, No. 1, Summer 1984, esp. pp. 149–150, 155, 181–186.

[33] Huth and Russett, 1988, p. 39.

[34] Matthew Kroenig, "Nuclear Superiority and the Balance of Resolve: Explaining Nuclear Crisis Outcomes," *International Organization*, Vol. 67, No. 1, January 2013, pp. 141–171.

[35] Paul Huth and Bruce Russett, "What Makes Deterrence Work? Cases from 1900 to 1980," *World Politics*, Vol. 36, No. 4, July 1984, pp. 496–526; Huth and Russett, 1988.

capabilities do not deter conventional attacks.[36] Many of these studies, moreover, focus on the role of nuclear weapons or superiority in determining crisis behavior rather than extended deterrence; there is no clear evidence that nuclear superiority determines extended deterrent outcomes.

Denial and Punishment: Hybrid or Complementary Strategies

The critical question is *how much* military force, especially in the local area of potential aggression, is enough to create a perception that the opportunity for an easy victory has been "denied." Both theoretical and empirical literature suggests that the answer need not be an unquestioned ability to "win." Instead, there is room for middle ground—deploying sufficient local forces to raise the cost of a potential attack, to create the inevitability of escalation, and to deny the possibility of a low-risk fait accompli without necessarily providing enough military power to unquestionably defeat the attack. Such a strategy would represent a hybrid of denial and punishment strategies and seek to raise the risk of aggression without promising defeat.

Even incomplete denial capabilities can create the risks of escalation, raising the potential that the aggressor would confront costs beyond those that the forces deployed to the local area can impose. The objective of such deployments is not an ability to win, but only to put up a fight so violent that it is likely to escalate.[37]

This approach was arguably the U.S. strategy in Europe during the Cold War. Glenn Snyder suggested in 1959 that the U.S. forces in Europe were "incapable of denying any territory to the Soviets that they wish[ed] to take with full force." The purpose of the U.S. presence was not in its inherent capacity for denial, "but rather in its indirect *complementary* effects—that is, in the extent to which it strengthen[ed] the probable or evident willingness of the West to activate the strategic airpower deterrent."[38]

Strong but not decisive local forces could achieve these results in several specific ways. First was a classic trip-wire function: Because of the presence of U.S. troops in the path of a potential Soviet advance, any attack on Germany would instantly involve war with the United States. Second, U.S. presence put "national honor and prestige" on the line in ways beyond mere verbal commitments. Third, a moderately strong defensive position would require the Soviets to deploy a massive force to ensure quick victory, affecting the cost of the action and having the side benefit of providing the West with more warning. Fourth, a significant U.S. force engaged in "a heroic delay-

---

[36] Vipin Narang, "What Does It Take to Deter? Regional Power Nuclear Postures and International Conflict," *Journal of Conflict Resolution*, Vol. 57, No. 3, June 2013, pp. 478–508. For similar findings, see Thérèse Delpech, *Nuclear Deterrence in the 21st Century: Lessons from the Cold War for a New Era of Strategic Piracy*, Santa Monica, Calif.: RAND Corporation, MG-1103-RC, 2012.

[37] As Snyder, 1959, p. 4, contends, "Even if our denial force were incapable of holding, the enemy would have to reckon that the stronger it is, the more likely we are to believe that the application of strategic airpower would be the marginal factor that would clinch victory"—thus encouraging escalation on our part.

[38] Snyder, 1959, pp. 8–10.

ing action" would provide the fuel for an "emotional mobilization that might well be the marginal factor which would persuade an indecisive leadership to take retaliatory actions." Finally, the stronger the force and the longer it could hold out, the greater the risk an aggressor would face of a accidental escalation to devastating consequences during the attack. Due to such functions, Snyder concludes, "forces beyond those necessary for the trip-wire and yet too weak to defend against a full-scale attack nevertheless do contribute to the deterrence of such an attack."[39]

In illustrating these concepts during the Cold War, Snyder laid out a scenario very much like the one the United States and NATO confront today in the Baltics. The Soviet Union could stage a fait accompli to grab some land, he suggested, and leave NATO with the risky choice of a counterattack to take it back. In the process, the Soviets would "do everything possible to inhibit a response," from deploying powerful units into the seized area to making public statements that they had no further ambitions. "They would offer to negotiate and perhaps hint that their occupation would only be temporary," while threatening massive escalation if NATO did indeed respond.[40]

Yet Snyder argued that forestalling such a scenario did not necessarily demand an outright capability to defeat the incursion. If NATO built reasonably strong defenses that could prevent a cost-free invasion, Snyder argued, "it would be committed in advance to a determined and costly defense of the territory, and the responsibility for deciding to set off a substantial conflict would rest with the enemy." The key was not so much the strength of the local forces, Snyder suggested, as the "automaticity" of the response.[41] To take just one example of actual policies that reflected such an approach, the Alert System that NATO maintained during the Cold War sought to achieve exactly this kind of automaticity by requiring member nations to make specific forces available to the Supreme Allied Commander Europe (SACEUR) when allies increased NATO alert levels in line with emerging threats. The Germans saw this as essential to ensure defense of Berlin and forward defense of the Federal Republic of Germany.[42]

One challenge with such hybrid approaches is that there can be no objective measurement of their sufficiency. The goal is to create enough local force to make a potential aggression so messy and violent that the defender will not be able to resist escalating. But that is a highly abstract and subjective measurement.

---

[39] Snyder, 1959, p. 9. Schelling, 1996, p. 112, seems to agree. "Forces that might seem to be quite 'inadequate' by ordinary tactical standards," he argues, "can serve a purpose, particularly if they can threaten to keep the situation in turmoil for some period of time. The important thing is to preclude a quick, clean Soviet victory that quiets things down in short order."

[40] Snyder, 1959, p. 23.

[41] Snyder, 1959, p. 23.

[42] One NATO planning document that has been declassified discusses the design and role of these alternate procedures. See North Atlantic Military Committee, *Memorandum for the Members of the Military Committee in Permanent Session: Study on Alert Measures in Support of Berlin Contingency Plans*, October 18, 1962.

At a minimum, we might suggest at least five conditions to measure the strength of the hybrid approaches Snyder is suggesting—to create an "escalatory trigger" that can convince an attacker it does not have an easy opportunity for a fait accompli.[43] First, the defender should take powerful political measures to reaffirm its commitment—public promises, legislative resolutions, alliance commitments, and diplomatic engagements. Second, the local forces should be of a type and character to signal an intention to powerfully confront any aggression (typically this means combined armed forces such as mechanized units). Third, those forces should be significant enough to impose serious costs on an aggressor's attack, and large enough so that such costs would cause a severe political reaction at home. Fourth, the local forces should be integrated with those of the local partner and employed with operational concepts that promise to bog down an attack and create an extended fight.[44] Fifth, the state conducting extended deterrence should have forces in the region capable of projecting power to assist the local units from the first hours of a conflict (even if that power will not be sufficient to win) in order to strengthen the impression that the conflict will quickly escalate.

This analysis points to a possible distinction in thinking about U.S. and NATO efforts to shore up conventional deterrence in Europe, and especially the Baltics, in the face of Russian threats. One approach would be to begin a long series of steps, ultimately including the deployment of significant new ground and air combat power in the Baltic States themselves, to achieve a local war-winning capability. A second and distinct approach would be to undertake improvements designed to introduce severe complications into Russian planning, denying both the prospect of a quick and easy victory and the hope that such a campaign could be kept isolated from a larger war with NATO. The second approach would not seek a full capability to win the local conflict—to defeat and push back any Russian attack—but would achieve improved deterrent effects through a combination of denial and punishment. We suggest some specific implications of this distinction in Chapter Four.

The Ability to Deny May Not Be Enough

One reason why favorable local force balances may not be sufficient for deterrence is that any effort to achieve such an advantage can run into the problem of subjective perceptions we discussed in Chapter One. Lebow has assessed historical cases in

---

[43] These criteria are not derived from specific empirical evidence. They are inferred from themes reflected in numerous studies examined for this analysis.

[44] These can include plans for unconventional warfare, widespread use of militias, concepts for an extended insurgency, and employment of passive defenses such as tank barriers and minefields. Emerging drone and swarm technologies offer opportunities for dramatic advances in such conceptual approaches. In this context, it is worth noting that all three Baltic States have plans for sustained unconventional warfare and civilian resistance following the collapse of conventional defenses, with the goal of making the cost of aggression and occupation very high and to buy time for NATO reinforcements to liberate captured territory.

detail and concluded that many aggressors provoked crises "without any good evidence that the adversary in question lacked the resolve to defend his commitment."[45] They *convinced* themselves that adversaries lacked resolve under the influence of motivated reasoning and because, for geostrategic or domestic political reasons, they had already decided that they *needed* to act. "When leaders believe in the necessity of challenging their adversaries," Lebow concludes, "they become predisposed to see their objectives as attainable." They become so "self-involved that they often fail to consider or, if they do, seriously distort the evidence available to them about how their adversary will respond."[46] Bruce Russett reviewed a number of deterrence cases and concluded that in at least five recent ones "the defender definitely had the ability to win any major conflict," and in two others "the defender had at least a marginal advantage." Even "clear superiority," he concludes, "provides no guarantee that his antagonist will be dissuaded."[47]

These dangers reinforce the lesson that states engaged in deterrence must pay close attention to the target state's beliefs and perceptions. They also indicate that even a capability to deny objectives may not be enough to prevent war, and point again to the need to evaluate aggressor motives and to draw in reassurances in deterrent strategies. If an overall strategy to avoid conflict does not address the underlying motivations for aggression, the deterrer may broadcast signals that are highly effective from the standpoint of classic deterrence theory—and still fail.

Perceived Willingness—or Obligation—to Fulfill the Deterrent Threat

Another factor that helps to determine success or failure is the subjective factor of willpower. Does a state have the resolve to fight for a claimed commitment? Deterrence depends in part on the perception of the "threatener's determination to fulfill the threat if need be" but, more important, on the potential aggressor's "conviction that the threat will be carried out."[48] Deterrence fails, Russett concludes, "when the attacker decides that the defender's threat is not likely to be fulfilled."[49]

Classic deterrence theory takes this issue of perceived willingness a step further, to perceived *obligation*. Schelling made much of the fact that, particularly in extended deterrence situations, a defender would often want to wriggle free of a commitment once challenged. Making a potential aggressor believe in the deterrer's commitment to respond thus became a central challenge of successful deterrence. Simply stating a commitment is not enough; a defender must make it clear that it *has no choice* but to

---

[45] Lebow, 1983, pp. 335–336.

[46] Lebow, 1987, pp. 198, 206.

[47] Russett, 1963, pp. 102, 107.

[48] Schelling, 1980, p. 11. The important thing is not merely having a capability; it is projecting the willingness, indeed the requirement, to use it; see Schelling, 1966, p. 36.

[49] Russett, 1963, p. 98.

react. This led to Schelling's analysis of threats that leave as little room for maneuver as possible—irrevocable commitments that bind a state to react, in part by creating significant costs for a defender if they retreat.[50] For Schelling, the ultimate goal of a commitment strategy was to maneuver one's self into a position where one has no choice but to respond.[51] The literature suggests a number of specific mechanisms for creating such unbreakable commitments.[52] These include:

- *Deploying trip-wire forces.* If American troops die in significant numbers, it will be difficult or impossible for a U.S. administration to back away from deterrent pledges.
- *Making public commitments that stake national prestige.* A deterring state can, as Schelling explains, "incur a political involvement" and thus "get a nation's honor, obligation, and diplomatic reputation committed to a response."[53]
- *Constructing basing infrastructure* in the country or area the deterrer wishes to protect, as a signal of ability and will to deny objectives.
- *Conducting exchanges and training programs* with the military of the country the deterrer wishes to protect.
- *Undertaking formal treaties* with countries that the deterrer will protect. Research suggests that formal alliances increase deterrence success in part by reaffirming in very public and difficult-to-escape ways a defender's resolve to fight.[54]
- *Selling or transferring arms and other militarily significant items to the country.*

An aggressor who is willing to make small moves toward its objective—the classic problem of "salami slicing"—can also test a defender's willingness to fulfill threats. In response, the deterring state must provide recurring signals of its commitment by punishing small aggressions.[55] Otherwise, small violations will snowball and overall deterrence could collapse. Yet there is also a dilemma: States cannot respond to every small slight or aggression around the world. A North Korean cyberattack against South Korea, or even a probing attack using maritime patrol craft, is an entirely differ-

---

[50] Schelling, 1980, pp. 24–27, 36, 131, 134, 137, 187–188. See also Russett, 1963, pp. 98, 100–101, which stresses that public commitments themselves are not sufficient.

[51] Schelling, 1966, pp. 43–44.

[52] Crawford, 2009, pp. 283–284.

[53] Schelling, 1966, p. 49.

[54] Huth and Russett, 1988, pp. 35, 40. There may be a distinction between "treaty allies" and countries that the United States has a more general legal obligation to support, but this has not been tested in the literature.

[55] Schelling, 1966, pp. 66–67; Schelling, 1980, pp. 41, 71.

ent matter from general war. Failure to respond to major actions could ruin deterrence; trying to respond to everything can "cheapen the currency."[56]

Perceived Degree of National Interest Engaged

Existing research on deterrence also points to the national interests of the deterring state—and, more important, the way they are perceived by a potential attacker—as a critical variable affecting deterrence success.[57] In brief, if a defender is seen to have vital interests at stake, a potential attacker will believe threats of response. If the interests appear secondary, an attacker can convince itself that the defender's threats are hollow.

In his early survey of extended deterrence cases, Bruce Russett placed significant emphasis on economic and political relations as the key to effective deterrence. Cases of successful deterrence tended to involve direct military assistance from the sponsoring power to the proxy it hoped to defend; strong economic ties; and intangible but still widely recognized cultural and social bonds between the two. Deterrence success is "heavily dependent on the tangible and intangible bonds between him and the lawn." Strengthening such bonds raises "the credibility of deterrence by increasing the loss one would suffer by not fulfilling the pledge."[58]

The interests can take various forms. Some studies have suggested that significant economic ties tend to be related to successful deterrence and that geographic proximity can also serve as a signal of interests.[59] More broadly, as suggested above, some studies indicate that the interests inherent in a region can be the dominant considerations in a response, as opposed to analysis of state-specific interests.

Ultimately, it is the *interaction* of perceived interests on the part of the attacker and defender that plays a significant role in determining deterrence outcomes. A perceived asymmetry of interests between the two can impact the effectiveness of extended deterrence.

The Reputation of the Deterring Power

The role of reputation and prior experience in enhancing or reducing the credibility of deterrence is hotly debated. Recent research suggests that the relationship of credibility to effective deterrence is highly contingent, and states can pursue credibility to counterproductive lengths.

In this context, the concept of reputation relates specifically to the perceived willingness to risk war to fulfill commitments. An initial phase of deterrence literature—as

---

[56] Schelling, 1966, p. 51.

[57] George and Smoke, 1974, p. 560. Jervis, 1989, pp. 314–317, distinguishes among three types of interest: intrinsic (the most powerful, such as the security of a state's own territory); strategic; and verbal, or commitment.

[58] Russett, 1963, pp. 103–107.

[59] Huth and Russett, 1984; Huth and Russett, 1988, p. 35.

well as the dominant assumptions of U.S. policymakers for most of the Cold War—made two broad claims: first, that a potential aggressor's assessment of a defender's reputation for resolve influences (sometimes dominantly so) its calculations of risk and cost, and second, that reputation accumulates through a nation's overall actions; thus, standing firm or backing down in crises totally unrelated to a given potential aggressor may still influence reputation. In other words, threats are interdependent: Threats fulfilled or abandoned in one place can have implications for other issues.[60] This belief produced dozens of statements from U.S. presidents justifying actions on secondary interests for reasons of credibility.[61]

A second generation of analysis, including numerous empirical studies trying to isolate the effect of credibility as a statistical variable explaining deterrence outcomes, came to a contrasting conclusion. Studies by Jonathan Mercer, Daryl Press, and others have suggested that leaders make situational, rather than dispositional, judgments about resolve and ask whether a possible defender would fulfill a commitment *in a specific case or context* rather than as a product of its character.[62] This research suggests that reputational commitments are not interdependent; failing to respond in one case does not necessarily have any bearing on an adversary's belief that a state will respond on other issues.[63] Quantitative research has also found little generalizable connection between overall reputation and deterrence outcomes.[64] In some cases, results have been complex—prior victory does not necessarily enhance reputation, for example, whereas previous defeats or examples of appeasement damage it.

A third generation of studies has produced a set of more discrete findings. One is that while reputation in general may not affect deterrence outcomes, previous and relatively recent interactions with the *same* potential adversary can affect calculations of risk and thus the possibility of aggression. Conciliation toward a specific potential

---

[60] Schelling, 1966, p. 55.

[61] If the United States did not react to North Korean aggression in 1950, it "would be an open invitation to new acts of aggression elsewhere." Harry Truman, quoted in Alex Weisiger and Keren Yarhi-Milo, "Revisiting Reputation: How Past Actions Matter in International Politics," *International Organization*, Vol. 69, No. 2, Spring 2015, p. 473.

[62] Jonathan Mercer, *Reputation and International Politics*, Ithaca, N.Y.: Cornell University Press, 1996, pp. 8–9. Mercer defines *reputation* as "a judgment of someone's character (or disposition) that is then used to predict or explain someone's future behavior" (p. 6). He further notes that only when a potential attacker works to "explain the target's behavior as a function of its character" and then acts based on that conclusion has reputation affected deterrence (p. 45). See also Daryl G. Press, *Calculating Credibility: How Leaders Assess Military Threats*, Ithaca, N.Y.: Cornell University Press, 2007. Similar conclusions are reached in Theodore G. Hopf, *Peripheral Visions: Deterrence Theory and American Foreign Policy in the Third World, 1965–1990*, Ann Arbor: University of Michigan Press, 1994. For further discussion, see Jervis, 1982–1983, pp. 8–13.

[63] For an argument that U.S. actions in Syria and Ukraine were not interdependent, see Peter Beinart, "The U.S. Doesn't Need to Prove Itself in Ukraine," *Atlantic*, May 5, 2014; see also the argument based on conversations with senior Russian leaders in Julia Ioffe, "How Russia Saw the 'Red Line' Crisis," *Atlantic*, March 11, 2016.

[64] See Huth, 1999, pp. 32–34, 41–43.

aggressor, therefore, could increase the chances that it would challenge deterrence at a later date.[65] Another set of recent research emphasizes the role of intrinsic interests in substantiating reputation; when potential aggressors perceive that a defender has major interests at stake, they are more likely to anticipate strong resolve.[66] Vesna Danilovic's work combines these two factors to some degree and suggests that the importance of a geographic region plays a critical role in the inherent credibility of threats to fight for specific states.[67]

The lens of perception—especially under the influence of motivated reasoning—will sometimes cause a potential aggressor to view a defender's resolve as the aggressor wants or needs it to be rather than as objective analysis would suggest it is. Especially in cases of extended deterrence on behalf of far-flung friends and allies, an aggressor can almost always convince itself that a distant defender will not have the resolve to fight. In cases where an aggressor is in the grip of such wishful, skewed thinking, even a very strong reputation for resolve might not dissuade attack.

The Seriousness of the Response Being Threatened

Finally, once a potential aggressor has assessed the credibility of the U.S. threat, it must evaluate the cost that the deterrer threatens to impose. If the proposed actions do not seem to impose a high cost, they will not be enough to deter.

This equation is entirely context-dependent. The sufficiency of a deterrent threat will depend in part on the value system of the aggressor and the degree of interests it has at stake in the disputed territory. An aggressor's opinion on the matter can also change over time if its own situation becomes more urgent or if the interests in a given area evolve.

This consideration therefore demands close attention as to what a specific potential aggressor values. It could be that a given threat could be very effective against one state and utterly futile against another: If the risk relates to global public opinion, for example, a democracy might be concerned whereas a dictatorship would not.

Variance in the potency of threats can extend even to the issue of local force balances and the whole concept of deterrence by denial. In some cases, the prospect that an attack would fail might not be enough to deter aggression. The outstanding modern

---

[65] Jonathan Shimshoni, *Israel and Conventional Deterrence: Border Warfare from 1953 to 1970*, Ithaca, N.Y.: Cornell University Press, 1988, pp. 231–234; Huth, 1999, pp. 41–43; Freedman, 2004, pp. 52–56; Long, 2008, pp. 14–15. Weisiger and Yarhi-Milo, 2015, make a more general argument to restore the validity of a general theory of reputation, but they acknowledge that their findings are significantly stronger within dyadic pairs—meaning that it would speak more to the issue of credibility with the same adversary over time rather than a more generic reputation for resolve.

[66] Daryl G. Press, "The Credibility of Power: Assessing Threats During the 'Appeasement' Crises of the 1930s," *International Security*, Vol. 29, No. 3, Winter 2004–2005, pp. 138, 140, 168–169.

[67] Vesna Danilovic, "The Sources of Threat Credibility in Extended Deterrence," *Journal of Conflict Resolution*, Vol. 45, No. 3, June 2001, 341–369.

example is Egypt in the 1973 Yom Kippur War. Egypt neither expected success nor required it to justify aggression; launching the attack was the perceived value. In other cases, an aggressor may have such powerful normative commitment to a course of action that the prospect of failure will not deter it from action; undertaking aggression has become a moral imperative.

Summary and Proposed Framework for Deterrence Effectiveness

This survey of existing research makes clear that many different variables influence deterrence outcomes. Their relationship to one another is not always clear, and the decisive combinations of variables can differ from case to case. Even superior military strength will not always suffice.[68]

Table 2.1 lays out the set of variables summarized in the preceding analysis. The goal of this framework is to identify all the factors associated—through detailed case analysis or empirical/quantitative studies—with success or failure specifically in the category of extended deterrence.

This set constitutes our initial framework, derived from the available literature. Chapter Three summarizes our quantitative and qualitative tests of the framework and offers a revised and improved version. Chapter Four then applies the revised framework to the Baltic deterrent relationship.

Summing up many of the considerations involved in the framework, Alexander George and Richard Smoke have offered an important formulation that speaks to the bottom-line measurement of the effectiveness of deterrence policies. They propose one essential variable as the potential aggressor's "judgment of whether the risks of a particular option open to him can be calculated and/or controlled so as to make that option an acceptable risk."[69] This phrasing speaks to the ultimate bottom line of a deterrence relationship: Does the attacker have a specific scheme it thinks it can get away with?

"In almost every historical case examined," George and Smoke write, "we found evidence that the initiator tried to satisfy himself before acting that the risks of a particular option he chose could be calculated and, perhaps more importantly, controlled by him so as to give his choice of action the character of a rationally calculated, acceptable risk."[70] This question presumes some answer to each of the three basic conditions outlined above. It assumes a significant enough degree of motivation so that the poten-

---

[68] Huth and Russett, 1988, p. 42, argue that "various kinds of influences . . . all affect the likelihood that deterrence will fail and/or deterrence crises escalate to war. None of these influences can be ignored, and a prudent decisionmaker will evaluate all of them. As a result, successful deterrence is a good deal more complicated than simply possessing strong military forces."

[69] George and Smoke, 1974, p. 523.

[70] George and Smoke, 1974, p. 527.

Table 2.1
Literature Review: Initial Set of Key Variables

Category	Variable
How intensely motivated is the aggressor?	1. General level of dissatisfaction with status quo and determination to create a new strategic situation.
	2. Degree of fear that the strategic situation is about to turn against the aggressor in decisive ways.
	3. Level of national interest involved in specific territory of concern.
	4. Urgent sense of desperation of requirement to act; whether aggressor is locked into course of action.
	5. Degree of aggressive, reckless, risk-accepting opportunism.
	6. Level of motivated reasoning in play; degree of wishful thinking, misperception of basic strategic context.
Is the defender clear and explicit regarding what it seeks to prevent and what actions it will take in response?	1. Precision in the type of aggression the defender seeks to prevent.
	2. Clarity in the actions that will be taken in the event of aggression.
	3. Forceful communication of these messages to outside audiences, especially potential aggressor(s).
	4. Timely response to warning with clarification of interests, threats.
Does the potential aggressor view the defender's threats as credible and intimidating?	1. Actual and perceived strength of the local military capability to deny the presumed objectives of the aggression.
	2. Degree of automaticity of defender response, including escalation to larger conflict.
	3. Degree of actual and perceived credibility of political commitment to fulfill deterrent threats.
	4. Degree of national interests engaged in state to be protected.
	5. Reputation for resolve with potential aggressor.
	6. Degree of threat posed to aggressor's values and interests by the specific responses threatened by defender.

tial aggressor is actively seeking opportunities for adventurism. It speaks to the clarity of interests to be protected, and actions to be taken, by the defender: Any calculation of the risks involved by the attacker must grapple with those issues. And it draws in the issue of communication of capability and will, which can exercise decisive influence on what an aggressor thinks it can accomplish.

But George and Smoke's specific language raises a critical issue not always captured in deterrence calculations: Does the potential aggressor see the risks as limited and controllable? Many of the specific variables outlined in Table 2.1 really add up to this fundamental judgment: Does the potential aggression represent a bounded risk with limited consequences, or an unbounded one with many unpredictable and highly perilous outcomes?

This specific condition points to an especially common route to deterrence failure: a clever stratagem chosen by an aggressor to allow it to follow its strong motivations to act even when the evidence suggests it will be highly escalatory. Many aggressors have fashioned complex schemes to manage risks; outstanding examples are plans by both the Soviet Union and the United States for major invasions (of Afghanistan and Iraq, respectively) whose costs and risks would be managed because they would only stay a short time and install favored exiles to run the country. If an aggressor can design for itself such an innovative (but often delusional) end run around the risks of aggression, the result will be to defeat the power of deterrent threats. We have therefore included one measurement in Table 2.1 to try to capture this specific route to conflict.

CHAPTER THREE

Evaluating and Revising the Framework: Quantitative and Case Study Assessments

As Chapter Two discussed, an extensive review of the available literature nominated 16 possible explanatory variables that could help assess the likely success or failure of a policy of extended deterrence of conventional aggression. We then tested that framework through a quantitative evaluation of extended deterrence cases since 1945, and then four in-depth case studies of extended deterrence success and failure. The complete analyses of those analyses appear in Appendixes A–E. This chapter summarizes the findings of those analyses and the lessons they hold for the 16 variables identified in the initial framework. Our goal was to confirm or disconfirm the significance of each variable through the quantitative and qualitative analyses summarized here.

As the chapter will discuss, the lessons of the quantitative and qualitative research are mutually supportive. They point to 12 of the originally identified 16 variables as being strongly confirmed. The chapter then offers the revised framework for effective extended deterrence of conventional aggression built on these dozen variables. This is the framework that the study will use when evaluating the health of extended deterrent relationships.

It is critical to be clear about the character of such a framework, and what it can (and cannot) accomplish. Our goal is not to predict deterrence failure or success but to offer a number of factors that are theoretically and empirically associated with those outcomes. This analysis aims for the result to be a guide to assess deterrent relationships, and to isolate areas of strength, weakness, and potentially enhanced deterrence. It is not a model, based on demonstrated, quantitatively specific causal relationships, that can generate specific forecasts.

Quantitative Assessment of U.S. Extended Deterrence Cases Since 1945

Appendix A offers a detailed overview of the quantitative methodology we employed in this study, presents some descriptive statistics of the findings of the case study analysis, and delves into a few variables and outlier results. In brief, in order to explore and apply the range of variables developed in the framework in Chapter Two, we analyzed and coded 39 cases since 1945 of U.S.-led extended deterrence—cases—that is, where the United States employed military and/or economic pressure (in addition to political pressure) *to deter territorial aggression*. We organized them by determining whether they were instances of "immediate" or "general" deterrence. General cases refer to U.S. efforts to deter long-term or broader threats, whereas immediate cases are U.S. efforts in response to an immediate threat and are often but not always subcases of a longer, general deterrence case. For example, the case of America's Cold War deterrent posture in Europe has four immediate deterrence subcases: the 1948 and 1961 Berlin crises, as well as the deployments of Jupiter and Pershing II missiles.

The diversity of cases—including geographic and temporal aspects, and which actors were involved—provided a broader lens through which to assess potential drivers of deterrence successes and failures. Of the 16 initial variables that the framework identifies, we coded five for the 39 selected cases:

1. How motivated was the aggressor?
2. How clear was the United States (both regarding what the United States wanted and the consequences the aggressor would confront if the United States did not get what it wanted)?
3. What was the local balance of forces?
4. What was the degree of U.S. interests involved?
5. Did the adversary believe the United States would respond?

We selected these five variables because they allowed us to represent each of the main categories in the framework with so-called proxy variables, while still making it plausible to code them. Our interpretation of the first and fifth variables, in particular, intends to capture some factors from the framework that would be difficult to code consistently across a number of cases; for example, the literature addresses the potential impact of motivated reasoning and wishful thinking, but these factors would be difficult to distinguish from regular or wholesale "judgment." Therefore, they are folded into a broader assessment of "aggressor motivation" and "belief the United States will respond."

We then coded whether deterrence succeeded or failed to look for patterns that could shed light on the question of "what deters and why." Four ongoing cases were not coded for success or failure, and so are excluded from the final tally of successful versus failed deterrence.

This effort suggested that U.S. extended deterrence has succeeded more often than it has failed. We determined there were 24 cases where U.S. deterrence efforts succeeded, and 11 cases where they did not.

The specific coding decisions connecting each case study with the variables were based on qualitative case research into each of the examples of extended deterrence. In many cases, the coding decisions were fairly binary and straightforward, such as where the United States had no clear deterrence statements at all (as in Iraq in 1990 or North

Korea in 1950). In other cases, though, where the evidence did not allow a clear and obvious judgment on the variable under consideration, the coding demanded a more nuanced judgment. We generally used a three-part scale—high, medium, and low—that allowed us to code the cases on a spectrum. Broadly speaking, cases that ended up having vague or conflicting evidence often ended up in the medium category, simply because there was no evidence to describe them as either high or low. However, as the results of the coding make clear, we did not judge it necessary to place the vast majority of cases into this middle ground; in many if not most instances, evidence was available to render a clearer judgment about whether a case did or did not reflect the variable.

In general, the case study findings align with those of the literature review and application of classic findings of previous deterrence research, as outlined in Chapters One and Two. For example, our analysis showed no deterrence failures when

- the aggressor's motivation is coded as low
- the United States is clear—both about what it wants, and the consequences for crossing the United States
- there is a clear U.S. and allied advantage in the local balance of forces.

Additionally, deterrence success (24 total cases) was highly associated with cases where similar criteria were met—times when the United States is clear (19 of 24 cases); U.S. interests were high (19 of 24 cases, with no success cases when U.S. interests are coded as low); and the adversary believed the United States would respond (15 of 24 cases) or was uncertain as to whether the United States would respond (nine of 24 cases). Deterrence failures, on the other hand, were associated with a highly motivated aggressor (nine of 11 cases); an ambiguous (five of 11 cases) or somewhat clear (six of 11 cases) U.S. commitment; or an ambiguous local balance of forces (eight of 11 cases).

There were several outlier cases, such as U.S. deterrence successes in the face of an adversary's advantage in the local balance of forces and U.S. deterrence failures when the aggressor believed the United States would respond. However, in general, the case study analysis supported the literature on when deterrence succeeds and why.

Qualitative Assessment: In-Depth Extended Deterrence Case Studies

Appendixes B–E present four detailed case studies of extended deterrence. We present herein a summary of each.[1] Taken together, they reinforce the lessons of the quantitative research, largely emphasizing similar variables.

---

[1] These summaries do not include the full citations that appear in this report's appendixes. Apart from direct quotes or references to specific U.S. policies, footnotes have been excluded from the summaries but are available in the relevant case studies in the appendixes.

The four cases were chosen to represent a combination of successful deterrence (Berlin and NATO's Northern Flank), failed deterrence (Iraq), and ambiguous outcomes in between (Georgia). We sought a varied set of cases that reflected different conditions for our key variables: some in which U.S. signaling and credibility had been very strong, and some in which they were weak; some representing a highly motivated aggressor, and some involving an aggressor without strong reasons to attack. Across all the cases we chose examples in which the local balance of forces was either negative or not strongly in U.S. favor, as such cases tend to bias the results in favor of deterrence success. We were interested, using the Baltic States as our ultimate parallel, in ways deterrence can succeed and fail in situations where the aggressor might believe it has a favorable local military balance.

In a few of these cases, our analysis considered the role of U.S. policy in affecting the choices of partners and allies. The cases suggest that part of the overall dissuasion effort in any complex case of extended deterrence will be shaping these choices to moderate the dynamics that could lead to conflict—or to directly restrain an ally from taking aggressive action. Such steps do not count as deterrence as classically understood—but in the narrower, pure definition of the term, they do involve the prevention of an unwanted action by another state.

Berlin

U.S. efforts to deter the Soviet Union from attacking West Berlin during the Cold War ultimately succeeded. Two acute crises did erupt over the city, in 1948 and 1961, in which the Soviet Union tested the credibility of U.S. commitments. These crises arose in large part due to Soviet insecurity over developments in Germany more broadly, but the choice of Berlin as a potential pressure point for the West also reflected Soviet perceptions that U.S. resolve to fight to stay in Berlin was at least initially uncertain. By the later Cold War period (1964–1989), after the demonstration of Western resolve in the two crises and the establishment of a modus vivendi between the two sides over Germany, the potential high costs and low benefits of further crises became clear and the Western position in the city persisted largely unchallenged.

The case of West Berlin has implications for each of the three main categories of variables affecting the success of deterrence in the framework developed in this report. From 1948 to 1963, the Soviet Union had relatively clear and strong motivations for ejecting the United States and its allies from Berlin, or at a minimum threatening to do so in pursuit of additional strategic imperatives. The Western enclave inside East Germany functioned as a threat to the Soviet satellite, providing a route for mass emigration and economic influence. It was also a convenient point from which to exert pressure on the United States to achieve Soviet goals in West Germany, first to arrest the formation of the West German state, and then to keep that state from developing nuclear weapons.

Despite these motivations, Soviet leaders never arrived at a decision to use even limited levels of force over Berlin. While they were willing to accept heightened levels of risk of inadvertent conflict to achieve their objectives, at no point did they intend to try to take the city by force and precipitate a war with the United States. While they may have ideally preferred to kick the West out of Berlin, proximate Soviet perceptions of U.S. commitments and capabilities shaped their goals. Soviet motivations to threaten the Western position in Berlin, although clearly affected by their perceptions of U.S. deterrence, should still be distinguished from other cases where adversaries had a clear intent and plan to initiate hostilities. Berlin therefore does not test the ability to deter adversaries that have become bent on attack, but it does demonstrate how deterrence can be used to prevent adversaries from considering such a course of action to begin with, despite clear incentives.

The clarity and credibility of U.S. threats to fight to stay in Berlin varied. In 1948, despite facing a United States that retained a nuclear monopoly, the Soviet Union assessed that it could simply cut access to the city, forcing the United States to withdraw and accede to Soviet demands, or itself be the one to initiate hostilities, a prospect the Soviets judged to be unlikely given the massive Soviet conventional superiority in Germany. At the time, senior U.S. officials seriously considered abandoning Berlin (although President Harry Truman was not among them). While the United States took steps to bolster the credibility of its willingness to go war if attacked in Berlin in 1948, such as the deployment of B-29 bombers to Europe, the Soviets never planned to force the crisis to the point of armed conflict. The unexpected success of the Western airlift in providing an alternative way out of the dilemma the Soviets had constructed helped to buy the United States a reprieve.

By 1961, despite substantial NATO investments, Soviet forces continued to dwarf NATO conventional forces. Meanwhile, the Soviets had developed their own nuclear deterrent, which, while still smaller and less easily deliverable than U.S. nuclear weapons, had become a substantial threat, including to the U.S. homeland. Against this backdrop Soviet premier Nikita Khrushchev assessed that whatever nuclear edge the United States might retain, President John F. Kennedy would employ it over a crisis in Berlin. While U.S. public statements regarding their determination to fight to retain their position in Berlin had remained clear, they were not initially credible to the Soviets. In part, this appears to have stemmed from Kennedy's refusal to intervene in the Bay of Pigs fiasco earlier that year, which Khrushchev had assessed as a sign of weakness. However, the lack of initial U.S. credibility also reflected the inherent difficulty of signaling U.S. willingness to risk nuclear war in order to maintain its position in an exposed enclave that it could not long defend conventionally.

Establishing this credibility required clear demonstrations of U.S. intent, including deployment of additional forces to Berlin and Western Europe during the crisis, as well as repeated, explicit public statements by Kennedy and other senior administration officials that conflict in Berlin would mean general war with the United States; the speakers would have suffered a devastating diplomatic and political cost had they later abandoned those statements. To be effective, U.S. threats did not need to guarantee a nuclear response to a conventional attack in the minds of the Soviet leadership; they had only to make such a response plausible enough, given the existential risks to the Soviets that would have been involved, to outweigh the relatively limited Soviet goals in Berlin, particularly after the Berlin Wall was built.

The less eventful success of deterrence in the later Cold War period can be tied to two factors. First, the enhanced credibility of U.S. deterrent promises was buttressed by the commitment demonstrated in 1961, as well as the additional investment of resources in the defense of Western Europe more broadly. Second, Soviet motivation to threaten the city lessened once the wall was constructed and the United States had tacitly promised to link the territorial status quo with pressure on West Germany to remain a nonnuclear state. Both factors were crucial in ensuring that deterrence continued to hold for the remainder of the Cold War.

The successful resolution of the 1948 and 1961 crises shows the necessity of clear, credible U.S. commitments for deterrence to work, but it also highlights the importance of tactical flexibility and recognition of the legitimate security concerns of the other party. In the 1961 crisis, in particular, the United States was able to preserve its position in Berlin and in Europe more broadly, but the Soviets did not come away empty-handed. U.S. acquiescence to the building of the wall, and more generally to granting the Soviets a free hand to run East Berlin as they saw fit, was seen at the time as a sign of weakness and lack of resolve that could embolden future Soviet aggression, for it represented a rollback of U.S. rights under postwar agreements.

However, Kennedy judged that Khrushchev's need to address migration flows through the city was acute, and Khrushchev's motivation to act to alter the status quo was strong. Kennedy was therefore willing to accept a weakening of U.S. rights in the city in order to allow the Soviets to stabilize the situation in a manner consistent with a continued Allied presence in West Berlin. Executing this limited retreat from previous U.S. positions while enhancing the credibility of U.S. promises to retreat no further required numerous signaling efforts, including explicit public commitments and military movements. In the end, though, this strategy (combined with the related understanding that the United States would prevent an independent West German nuclear capability) proved effective in limiting Soviet motivations to risk war again over Berlin. At the same time, it maintained the clarity and credibility of U.S. commitments to fight over the city.

This case therefore highlights the types of steps that may be required for the United States to make clear and credible commitments in the service of extended deterrence. While local conventional military superiority was not required in this case, a credible willingness to escalate to general war was. Establishing this willingness required costly signals of political and military commitments on the part of the United States. At the same time, a blind refusal to consider any modifications to previous commitments would likely have increased the risk of deterrence failure and conflict in 1961 given the security concerns the Soviet Union faced. The case demonstrates that efforts to ensure the clarity and credibility of U.S. commitments also need to be considered in light of the effect that they may have on adversary motivations.

Deterring Saddam, 1990

President George H. W. Bush inherited his approach to the Baghdad regime from the administration of President Ronald Reagan, which viewed Saddam Hussein as the bulwark preventing the region from falling under the influence of the newly established Islamic Republic of Iran. As such, the United States restored relations with Iraq in 1984 (Iraq had cut ties in 1967 following U.S. support for Israel during the Six-Day War) and tilted toward Saddam during most of the Iran-Iraq War.

In October 1989 the Bush administration issued National Security Directive 26 (NSD 26), which would serve as the guideline for U.S. policy in the Persian Gulf until Saddam's invasion of Kuwait.[2] The directive claimed that normalized relations with Iraq were in the U.S. national interest. To maintain relations with Iraq, the United States would need to moderate Saddam's behavior by providing Iraq with economic incentives and creating opportunities for U.S. firms to help in Iraq's postwar reconstruction. According to then–Deputy National Security Adviser Robert Gates, U.S. officials did not expect Saddam would change dramatically; they did hope, however, that he could become a more predictable dictator like Syria's Hafez al-Assad. Meanwhile, NSD 26 also declared that U.S. access to Persian Gulf oil and preserving the security of friendly regional states were vital to its national security, and that Saddam should be made to understand that any further use of chemical or biological weapons would result in sanctions.

Projecting their frame of mind onto Saddam, most U.S. officials failed to comprehend fully the threat Iraq posed to Kuwait. They assumed that Saddam would refrain from aggressive behavior and focus instead on reconstructing his country in the wake of a costly war with Iran. But while Iraq had cut its forces by half following its cease-fire with Iran, Saddam's 400,000-man army was still the largest in the region. Furthermore, the widespread belief that "Arab countries did not invade other Arab countries" blinded Washington to the possibility of an Iraqi invasion.[3] Therefore, all of Iraq's aggressive actions and rhetoric were interpreted merely as bluffs to gain concessions from Kuwait.

Because of the Bush administration's faulty assumptions, the United States did not issue clear and strong warnings to Saddam regarding the costs he would incur should he invade Kuwait. Bush's mixed messages to Saddam are the most commonly cited factor in explaining Washington's failure to deter the Iraqis. Several months after the Iraqi invasion, even Secretary of State James Baker contended that the occupation could "absolutely" have been prevented if the United States had issued strong warnings to Saddam.

In retrospect, the Bush administration did not employ methods that tend to make for an effective deterrence strategy. Since the Gulf War, it has often been argued that

---

[2] National Security Directive 26, October 2, 1989.

[3] Michael R. Gordon and Bernard E. Trainor, *The Generals' War: The Inside Story of the Conflict in the Gulf*, New York: Little, Brown and Company, 1995, p. 5.

the invasion could have been prevented if the Bush administration had more directly communicated what specific costs the Iraqis would incur if they attacked Kuwait. Perhaps, for instance, the United States should have stationed troops in Kuwait to indicate its capability to push back an Iraqi invasion. Bush also could have issued clearer public statements to stress his commitment to Kuwaiti security.

However, it is important to ask whether Saddam could even have been deterred in the first place. As Saddam so highly valued regime survival, he should have been deterrable at some point in time. While he was willing to incur U.S. retaliation, he was not suicidal. For instance, he refrained from using weapons of mass destruction against U.S. forces because he and his advisers were certain that Washington would respond with nuclear weapons.[4]

By summer 1990—when Saddam's rhetoric and actions led at least a few U.S. intelligence analysts to alert officials to the possibility of an invasion—multiple factors existed that would have posed great obstacles to any attempt at deterrence. While Saddam had been paranoid about the United States as far back as the 1986 Iran-Contra Affair, by 1990 he had become certain that Washington and its allies were trying to overthrow him. Therefore, any attempts to deter Saddam by sending forceful messages could have simply convinced him that an attack was even more imminent.

Furthermore, as the economic situation in Iraq worsened, the potential of domestic unrest increased the cost to Saddam of doing nothing—in his eyes meaning not retaliating against the U.S.-Kuwaiti conspiracy. And once he had mobilized his Republican Guard in Basra on July 21, backing down in the face of U.S. pressure would have seemed even costlier. He appears to have had some belief that the United States was by this time conspiring with Kuwait against him to destabilize his regime. Perhaps in the early months of 1990, more concerted U.S. attempts to persuade the Kuwaitis to make concessions could have allayed Saddam's fears, but by the summer, any U.S. diplomacy may have been interpreted as an attempt to lull Saddam into complacency. Only a complete Kuwaiti capitulation to Iraqi demands would have staved off an invasion at that point.

At the same time, there is no guarantee that Saddam would not have been provoked to aggression by another alleged conspiracy. According to Iraqi foreign minister Tariq Aziz, Saddam believed that "Iraq was designated by George Bush for destruction, with or without Kuwait."[5] Therefore, at least by invading Kuwait, Saddam would have a bargaining chip. In the words of Janice Gross Stein, "Once Saddam concluded that the United States was determined to undermine his regime, reassurance and deterrence became virtually impossible, even had the United States clearly defined its commitments and consistently communicated its benign intentions."[6]

---

[4] "Oral History: Wafic Al Samarrai." *PBS Frontline*, 1995.

[5] "Oral History: Tariq Aziz." *PBS Frontline*, 1995.

[6] Stein, 1993, p. 135.

Stronger efforts at deterrence may have prevented Saddam from *fully occupying* Kuwait, however. Stationing U.S. forces in Kuwait could have served as a trip wire, preventing the Iraqis from thinking that an occupation would be a fait accompli. After his capture in 2003, Saddam told his American captors that he would not have attacked Kuwait had he realized the level of force with which the United States would respond. Yet this claim contradicts statements made directly after the Gulf War by several senior Iraqi officials and discounts Saddam's heightened level of paranoia at the time. Therefore, it is probably more accurate to say that Saddam would not have occupied Kuwait had he been made aware of the cost.

The U.S. failure to deter Saddam Hussein presents several implications for this report's deterrence framework. Washington was unable to deter Iraq because it lacked a clear understanding of the geopolitical context in the Middle East. Iraq and the United States had been de facto allies in Saddam's war against Iran, but the Iraqi dictator believed the U.S. government to be an enemy intent on his overthrow. In addition, Saddam felt a deep sense of grievance against the Arab monarchies of the Persian Gulf; Iraq had served as a bulwark against Iranian expansion, only for its Arab brothers to betray it. Saddam also viewed himself as a historical leader destined to unite the Arab world under his authority. But the United States remained largely ignorant of these realities.

The lack of U.S. understanding regarding Iraq's intentions and regional geopolitics undermined Washington's efforts to shape Saddam's thinking. Positive inducements and relatively vague threats of punishment failed to deter Saddam's occupation of his smaller and much weaker neighbor. The Iraqi regime knew that the Kuwaiti military was no match for its war-hardened military machine. Iran was a weakened regional power while Saudi Arabia was dependent on the United States. Only the United States—the world's only remaining superpower—could prevent an easy Iraqi conquest of Kuwait. The absence of major U.S. forces in the region, a perceived lack of U.S. resolve to defend Kuwait, and the overall mixed—if not confusing—signals from Washington appear to have facilitated Saddam's decision to occupy.

Moreover, Iraq's invasion of Kuwait was not the result of a single decision point, but rather determined by a number of circumstances, including Saddam's belief that he had little choice but to take action against Kuwait in order to survive economically and politically. The authoritarian nature of the Iraqi government and Saddam's paranoid and brutal style of rule only reinforced his decision to invade.

NATO's Northern Flank in the Cold War

During the Cold War, NATO's Northern Flank comprised Denmark and Norway; it was lightly defended, removed from the Central Front, and vulnerable to potential military aggression from the east. In case of an attack, the two countries' militaries would have had difficulty holding out until allied reinforcements arrived. The governments of Denmark and Norway further imposed severe limitations on their participation in the NATO alliance by banning foreign military personnel, bases, and nuclear weapons from their territory. The overwhelming Soviet military presence just across the USSR's shared border with Norway—comprising naval, air, ground, and nuclear forces—compounded the situation.

The situation of the Northern Flank countries throughout the entire period of the Cold War presents a unique case of U.S. extended deterrence. The United States did not begin to invest serious efforts into deterring the Soviet Union in this region until the last decade of the Cold War, but the Soviets did not undertake military aggression against Denmark or Norway in the meantime. U.S. deterrence efforts here ultimately succeeded due to a combination of the Soviets' own limited objectives for the region, the clarity of the U.S. deterrence messaging, and the aggressiveness of U.S. and NATO deterrence.

Though both the United States and Soviet Union quickly recognized the strategic value of the Northern Flank, for most of the Cold War this region was not the site of major contestation between U.S./NATO and Soviet/Warsaw Pact forces. A confluence of factors ensured this stability—and, most important, the perceptions that the various players had of the regional security situation. Finland signed the treaty of Friendship, Cooperation, and Mutual Assistance with the USSR on April 6, 1948, thereby guaranteeing its own neutrality and its status as a reliable buffer state for the Soviets. Sweden, having managed to avoid occupation during World War II, decided to maintain armed neutrality in the postwar period, creating yet another buffer between the East and the West.

Both Denmark and Norway decided to abandon their long-standing traditions of neutrality and join NATO because of the experience of having been invaded and occupied by Nazi Germany. At the same time, however, they remained wary of antagonizing the Soviet Union and so pursued a strategy of simultaneous deterrence and reassurance. Accordingly, they joined NATO to have as an ultimate security guarantee, but they simultaneously imposed a number of important restrictions on their participation in the Alliance.

There were moments of alarm; for example, in 1979 when the Soviet invasion of Afghanistan and internal exile of Nobel Peace Prize winner Andrei Sakharov provoked strong reactions among the Norwegian public. But a Soviet attack against the Northern Flank never materialized. U.S. deterrence in this case appears to have succeeded first because the Soviet Union held limited objectives, and second because a gradual shift in U.S. and NATO strategic thinking about northern Europe from the late 1960s onward caused the United States and its allies to become more assertive and proactive in their defense of the region.

U.S. deterrence efforts on NATO's Northern Flank during the Cold War show that the potential aggressor's level of motivation contribute greatly to deterrence success or failure. The U.S. deterrence messaging—regarding NATO and, more specifically, Denmark and Norway—was very clear. Successive U.S. administrations were committed to the defense of Western Europe, but more concrete U.S. efforts to defend the Northern Flank in particular—such as a forward naval posture and large-scale exercises—did not begin to materialize until the late 1970s and early 1980s. In the interim, U.S. and NATO planners focused on the Central Front.

During this time the Soviet Union could have taken the opportunity to launch an attack and undermine U.S. extended deterrence. Ultimately, the Soviet Union's limited objectives of denial, rather than possession, and the Soviets' resulting low level of motivation to initiate an attack against Denmark or Norway, prevented military aggression.

The question remains as to whether the United States would have achieved the same success had it had strengthened its deterrent message and become more proactive in northern Europe ahead of the 1980s. There are indications that U.S. extended deterrence might actually have come closer to failing if it had undertaken this shift earlier. Northern Norway, adjacent to the Kola Peninsula and the headquarters of the USSR's Northern Fleet, was an area of high strategic importance to the Soviet Union. Its loss could have seriously threatened the Soviet Union's naval capabilities, as well as its nuclear deterrent.

Before the 1980s the United States did not seriously threaten the prevailing security situation in the area because its attention was focused elsewhere. This focus elsewhere likely contributed to the Soviet Union's limited objectives (denial rather than possession) in the region. The United States increased its forward presence during the 1980s, when the Soviets became aware of and eventually accepted (albeit reluctantly) the significant and widening technological gap between the two countries' militaries. If the United States had been more proactive earlier, it might have increased pressure on the Soviets, but it also would have risked inflaming a sense of desperation; the Soviets might have come to believe they had to act before the strategic situation turned irrevocably against them in this strategically important region. The still relatively narrow capabilities gap between NATO and the Warsaw Pact militaries would have risked a prolonged and risky confrontation.

The Northern Flank case appears to suggest that, in areas where the potential aggressor has low motivation to attack, a more low-key approach—combined with active vigilance—might be sufficient to deter. A more proactive posture, on other hand, might work if the defender has an overwhelming advantage over the potential aggressor—and the aggressor is aware of this fact. Otherwise, a proactive posture might push the potential aggressor into action, causing the very act that the defender had sought to prevent in the first place.

Russian Aggression Against Georgia

The weeks and months that led to the so-called Five-Day War (August 7–12, 2008) between Russia and Georgia, as well as the war itself, present two cases of deterrence involving the United States.[7] First, the United States attempted to prevent Georgia from responding militarily to provocations from South Ossetian separatists—a

---

[7] While the European Union also played a role in these deterrent efforts (particularly the second one), this chapter focuses primarily on deterrence efforts by the United States, consistent with other case studies in this report.

response that, it was thought, would automatically trigger military retaliation from Russia. Second, the United States tried to prevent Russia from crushing Georgia militarily, at a time when Russian forces had crossed into Georgia's undisputed territory and were advancing toward Tbilisi. U.S. deterrent efforts were successful in the latter case but not the former, suggesting it is sometimes easier to convince a rival (Russia) than a friend (Georgia). This case of successful deterrence, however, owes more to Russia's lack of interest in taking over Tbilisi than to clear messaging or powerful threats on the part of the United States.

As tensions built up between Russia and Georgia—first around Abkhazia, and then around South Ossetia—the United States tried to deter its Georgian ally from intervening in the breakaway provinces—a move the administration of President George W. Bush believed would provide Russia with a pretext to respond forcefully. In her memoir, then–Secretary of State Condoleezza Rice recalls about Mikheil Saakashvili that "we all worried that he might allow Moscow to provoke him to use force."[8] U.S. efforts to prevent Saakashvili from militarily confronting Russia in South Ossetia represent a case of failed deterrence. On August 7, 2008, Georgian forces shelled Tskhinvali, prompting Russia to claim that its nationals—whether Russian peacekeepers or South Ossetians with Russian citizenship—required protection and thus causing the military invasion of South Ossetia before advancing into Georgia.

If Saakashvili did believe that war with Russia was highly likely or even already underway, his options outside a military attack were indeed limited. Abkhazia and South Ossetia figured prominently in the 2004 campaign that had gotten him elected, and on his agenda; he was unlikely to survive politically if Russia took over South Ossetia. Even without taking into account the political salience of South Ossetia, not responding in kind to a Russian attack would have inflicted damage beyond repair to Saakashvili's credibility at a time when his crackdown on the peaceful protests of November 2007 and subsequent state of emergency in Tbilisi had eroded his popularity. Finally, if the Russian attack did not stop in South Ossetia but continued instead toward Tbilisi (as it eventually did), it would almost certainly have removed Saakashvili from power. As Vano Merabishvili, then Georgia's minister of the interior, put it in an interview after the fact,

> We were faced with a situation where there was no choice. Or are you saying I had to stand by in Tbilisi and wait for the Russian tanks? Maybe we gained some time by acting fast. What would have happened if the Russian tanks invaded, and without resistance got to Tbilisi?[9]

---

[8] Condoleezza Rice, *No Higher Honor: A Memoir of My Years in Washington*, New York: Crown Publishers, 2011, p. 685.

[9] Vano Merabishvili, interview, October 29, 2010, translated from Russian by Samuel Charap. Svante E. Cornell, "War in Georgia, Jitters All Around," *Current History*, October 2008, p. 312, argues that Georgia's use of force did slow down the progression of the Russian advance, possibly giving more time for negotiations to succeed before the Russian forces could reach Tbilisi.

Georgia's motivation to launch its attack in South Ossetia was therefore extremely strong, since all other options would have led to riskier or worse outcomes for Saakashvili. As Ronald Asmus puts it, "The Georgian decision to use force was made at the last second by a leader who felt cornered."[10]

In their memoirs, U.S. officials who communicated with Saakashvili as tensions with Russia mounted emphasize their efforts to convey to the Georgian leader that he should not use force in the breakaway regions, and that if he did, he should not expect military support from the United States. Condoleezza Rice, for instance, recalls,

> Finally I thought I'd better get tougher. "Mr. President, whatever you do, don't let the Russians provoke you. You remember when President Bush said that Moscow would try to get you to do something stupid. And don't engage Russian military forces. No one will come to your aid, and you will lose," I said sternly. He got the point, looking as if he'd just lost his last friend. I tried to soften what I'd said by repeating our pledge to defend Georgia's territorial integrity—with words. He asked if I'd say so publicly. I did, avoiding any language that might be misinterpreted as committing us to Georgia's defense with arms.[11]

This recollection, if accurate, suggests Saakashvili understood clearly that the United States would not come to his help in case of a confrontation with Russia. Then–National Security Advisor Stephen J. Hadley similarly remembers, "We made all kinds of signals to Putin to stay out and the president made all kinds of signals to President Saakashvili not to provoke Putin. I remember he said, 'Don't provoke Putin. You can't handle him and we will not be able to save you from Putin.'"[12] According to one journalistic account, President Bush took Saakashvili aside during the NATO Summit in Bucharest to tell him, "The U.S. would not start World War Three on his behalf."[13]

Based on interviews with the White House chief of staff and the U.S. ambassador to Georgia at the time, Angela Stent notes that Bush had "explicitly warned Saakashvili not to let the Russians provoke him and not to use force to take back the regions, making it clear that the United States would not come to Georgia's rescue if it did."[14] The message that Georgia should not go into South Ossetia was repeated by

---

[10] Ronald D. Asmus, *A Little War That Shook the World: Georgia, Russia, and the Future of the West*, New York: St. Martin's Press, 2010, p. 49.

[11] Rice, 2011, p. 686.

[12] Elise Labott, "Stephen J. Hadley Looks Back on 9/11, Iraq, and Afghanistan," October 22, 2014, Council on Foreign Relations.

[13] Andrew Cockburn, "Game On: East vs. West, Again," *Harper's*, January 2015.

[14] Angela Stent, *The Limits of Partnership: U.S.-Russian Relations in the Twenty-First Century*, Princeton, N.J.: Princeton University Press, 2014, p. 168.

then–Assistant Secretary of State for European and Eurasian Affairs Daniel Fried to Georgia's foreign minister Eka Tkeshelashvili as late as August 6, 2008.[15]

While a few accounts dispute the clarity of the U.S. message to the Georgian leadership, Asmus's well-documented account of the war concludes,

> Speculation over whether Washington had given Tbilisi some kind of green light misses the point. No senior Georgian official has actually ever suggested that Washington did so. On the contrary, they all admit that warnings had been given repeatedly by senior American and European officials.[16]

This case suggests that the framework outlined in Chapter Two is applicable to instances of countries deterring allies, rather than just rivals or enemies, from pursuing a certain course of action. Most of the variables highlighted in Chapter Two as relevant for deterrence were present at a high level in this case. Georgia's motivation to attack Tskhinvali was very high; the U.S. message was very clear and reasonably credible. Deterrence failed because the first factor—Georgia's motivation—was so strong that the Georgian leadership accepted all potential costs. The variable that plays a fundamental role in this dynamic is Tbilisi's belief that it was locked into a course of action. While other variables pertaining to the aggressor's motivation can be balanced against the costs of aggression, costs become largely irrelevant if the aggressor believes that there is no other option besides aggression.

U.S. attempts at stopping Russia's advance toward Tbilisi after its military forces routed Georgia's represent a second case of deterrence. At first glance, this case seems successful. The United States indicated to Russia that pursuing its offensive would be costly, and Russia stopped. Yet successful deterrence would have required Russia to have had the intention to take Tbilisi, and to have modified its plans specifically in reaction to the U.S. message in order to avoid the costs the United States had threatened to impose.

Yet a closer look suggests instead that Russia's motivation to reach Tbilisi and possibly remove Saakashvili from power was low, and that it did not see the U.S. threat as credible—although evidence on this second point is more limited. U.S. efforts at stopping Russia's advance are therefore not a case of successful deterrence. Nor, however, are they a case of failed deterrence—that would have entailed Russia *not* stopping its advance in response to the U.S. message. Rather, they are a case where the defender may have believed it was deterring an aggressor effectively while, in reality, there was little to deter.

---

[15] Helen Cooper, C. J. Chivers, and Clifford J. Levy, "U.S. Watched as a Squabble Turned into a Showdown," *New York Times*, August 17, 2008.

[16] Asmus, 2010, pp. 30–31.

This second case presents an almost mirror image of the first one regarding the key variables outlined in Chapter Two. While for U.S. efforts to deter Georgia most variables were at a "high" level, in the case of U.S. efforts to deter Russia they were all on "low." The fact that Moscow initially wanted to get rid of Saakashvili, but ended up not pushing its advantage when that objective was near (or at least closer than Moscow had ever been), suggests that Russia kept its objectives flexible and made an opportunistic decision not to take Tbilisi.

The United States chose to play a secondary role in the handling of the crisis in order to avoid a possible confrontation with Moscow, which explains why the U.S. message did not attempt to convey any clear or powerful threat. This points to the role that deterrence can play in crisis escalation—a role that was well understood by the Bush administration—as threats made, particularly military ones, have either to be carried out, with the potential of meeting a response in kind, or risk being "empty threats," with broad international and domestic implications.

The two cases of deterrence discussed in this section show a divergence between the quality of the deterrent message and the outcome of the deterrence effort. In the first case, the United States clearly expressed a credible threat message to Georgia, but Georgia went ahead with its attack against Tskhinvali anyway. In the second case, the U.S. deterrent message to Russia was weak by any measure, yet Russia did stop its advance toward Tbilisi, as the United States had hoped. In both cases, the factor that proved of critical importance to predict failure or success of deterrence was the degree of motivation of the aggressor. Georgia felt locked in a course of action, found itself with no good options, and went ahead with the attack regardless of what price it might have to pay for this decision. Moscow likely found that enough of its strategic and tactical objectives had been achieved by August 12, and that pushing into Tbilisi would have been more trouble than it was worth.

The fact that the degree of aggressor's motivation was of paramount importance also means that in both cases the outcome of U.S. deterrent efforts owed little to U.S. actions. Yet this does not mean that this outcome was entirely outside Washington's control. For instance, the United States could have taken some diplomatic steps earlier, as tensions escalated between Russia and Georgia in the months and years that preceded the August 2008 crisis. Extended deterrence against Russia at the time could have prevented Georgia from eventually finding itself in a situation where it saw a Russian military intervention as inevitable and military action as the only option. Extended deterrence would have come with its own costs, however, as it would have likely made U.S.-Russia discussions more difficult or conflictual on issues—such as missile defense—of greater strategic importance to the United States than the fate of Abkhazia and South Ossetia.

Unsurprisingly, it was easier for the United States to provide a clear and convincing deterrent message to its Georgian ally than to Russia. Saakashvili had been courting U.S. support, which he desperately needed to achieve his objective of setting his country on a westward course and joining NATO. No matter what actions the United States did or did not take, it would have been unlikely to alienate Georgia—and even if it had, the strategic consequences would have been minimal. Russia was a different story, and Washington was careful to steer away from anything that might look like, or trigger, a confrontation. While many of President Bush's decisions (particularly on missile defense) antagonized Moscow, there were efforts—including as recently as April 2008, in Sochi—to reinvigorate cooperation between the United States and Russia. As a result, Washington was exceedingly cautious in its messaging to Moscow. This prevented the United States from issuing a clear deterrent message and making threats that it did not want to deliver on.

Overall Lessons: Qualitative Case Studies

The qualitative cases point to a handful of especially important lessons. One was the critical role of a "firm but flexible" approach as described in the deterrence literature—mixing accommodation with firmness to deter attacks while meeting enough of a potential adversary's interests to keep it from attacking out of desperation. This factor emerged in the Berlin and Nordic cases, in particular; its absence helps to explain the failure of deterrence in Kuwait in 1990. It points to the conclusion that successful deterrence is not all about threats—it is also about reassurance, even of a potential aggressor.

A second primary theme that emerges from the case studies is the simple but indispensable role of clarity. Again and again, aggression occurs when a potential attacker—such as Saddam Hussein in 1990 (or, in a case not examined here in depth, Joseph Stalin and Kim Il-sung in 1950)—is unsure about the potential U.S. reaction. Cognitive factors magnify this risk: Aggressors engage in wishful thinking and will jump at excuses to believe that the United States does not have the willpower to respond as it may have threatened. In order to counteract the power of intention and wishful thinking, the United States must be very clear about what it will respond to—and support that clarity with diplomatic and military steps to convince an aggressor that it has bound itself to responding.

The cases also suggested that the degree of aggressor motivation is an absolutely critical variable governing deterrence success. Weakly motivated aggressors—those who see little positive risk-reward calculus in aggression—will not challenge even fairly obvious vulnerabilities—as in the Soviet Union's restraint with regard to Norway in the Cold War. On the other hand, aggressors in the grip of a powerful imperative will risk profound geopolitical consequences and confront very distinct potential for military loss—as in Iraq in 1990, and other relevant cases, such as the Soviet Union and Afghanistan in 1979 and Egypt in 1973.

An intensely motivated aggressor, in other words, can be almost impossible to deter. Such intense motivations, however, typically come from strong threat perceptions—which can be the product of a failed management of the "firm but flexible" component of deterrence. A major requirement for effective deterrence is therefore not merely—or at all—making violent threats against a potential aggressor, but rather *managing their threat profile and perception of risks and opportunities* so that they do not get to the point of seeing no alternative to war.

On the other hand, this analysis does not place such essential emphasis on a factor commonly believed to determine deterrence outcomes: the local balance of military forces. The empirical literature refers to a favorable balance as a factor likely to enhance deterrence, and our own quantitative analysis agrees that favorable local and more general balances of military power are associated with deterrence success. But our research also argues against *equating* deterrence and the local balance of forces, as is sometimes done, or concluding that such a positive balance is either necessary for deterrence (that deterrence will nearly always fail without it) or sufficient for deterrence (that deterrence will always or nearly always succeed with it). Both the quantitative and qualitative analyses agree that deterrence can succeed *without* a favorable local balance and can fail *with* such a favorable balance.

Local military balances therefore emerge as an important but not always decisive factor in determining deterrence success or failure. In some cases where the local balance was highly unfavorable, such as in Berlin and the Nordic countries, deterrence still held. Other gaps and vulnerabilities, as in Iraq and Kuwait in 1990, proved more dangerous. Although they have not been examined in detail in this study, other cases that have been treated at length elsewhere demonstrate the other side of the coin: that some aggressors are not deterred even when confronted with a significant possibility or even probability of defeat. Worryingly, this is true even when the prospect of defeat is discussed at length among their senior leadership, many of whom believed they were likely to lose the war. This pattern is evident at least in part in Japan's decision for war in 1941, in Germany's choice to invade the Soviet Union that same year, and in the Soviet Union's intervention in Afghanistan in 1979.

Our quantitative and qualitative analysis therefore suggests that the local military balance can exercise an important signaling function, but usually only in combination with other factors. While an ability to prevail in the local fight is *useful* for extended deterrence—and an unquestioned ability to win can be among the most effective means of deterrence—such a capability is in many cases neither necessary nor sufficient on its own for extended deterrence success.

A Revised Framework of Factors Associated with Deterrence Success

Taking the analysis of the quantitative and qualitative cases together, then, we can confirm 12 of the original 16 variables as helping to determine the success of extended deterrence of major conventional aggression. The revised framework appears in Table 3.1.

Table 3.1
Key Variables

Category	Variable
How intensely motivated is the aggressor?	1. General level of dissatisfaction with status quo and determination to create a new strategic situation.
	2. Degree of fear that the strategic situation is about to turn against the aggressor in decisive ways.
	3. Level of national interest involved in specific territory of concern.
	4. Urgent sense of desperation, need to act.
Is the defender clear and explicit regarding what it seeks to prevent and what actions it will take in response?	1. Precision and consistency in the type of aggression the defender seeks to prevent.
	2. Clarity and consistency in the actions that will be taken in the event of aggression.
	3. Forceful communication of these messages to outside audiences, especially potential aggressor(s).
	4. Timely response to warning with clarification of interests, threats.
Does the potential aggressor view the defender's threats as credible and intimidating?	1. Actual and perceived strength of the local military capability to deny the presumed objectives of the aggression.
	2. Degree of automaticity of defender response, including escalation to larger conflict.
	3. Degree of actual and perceived credibility of political commitment to fulfill deterrent threats.
	4. Degree of national interests engaged in state to be protected.

We do not conclude that all 12 need to be present in order for deterrence to succeed. The importance of these factors can and will vary by circumstances. But if the United States wants to assess the health of a deterrence relationship, the literature on deterrence, as well as our own efforts to assess its findings, suggests that these 12 factors are a good starting point. Chapter Four applies these factors to one case study as an example of such an assessment: the deterrence of Russian conventional aggression against the Baltics.

CHAPTER FOUR
Applying the Revised Framework: Deterring Russia in the Baltic Region

Russia's 2014 annexation of Crimea and its ongoing military operations in the east of Ukraine serve as a reminder—after the war in Georgia in 2008—that Russian president Vladimir Putin does not see established European borders as inviolable, and is willing to intervene militarily to protect what he perceives to be Russia's national interest. The Baltic States of Estonia, Latvia, and Lithuania, which joined NATO in 2004, have since routinely been mentioned as potential targets of future Russian military adventurism based on their geographic proximity to Russia; the fact that Estonia and Latvia have sizable Russian minorities; a history of diplomatic and military tensions; and their standing in the way of a physical connection between Russia and its enclave of Kaliningrad.[1]

The Baltic States have historically lived under the threat of invasion or annexation from their powerful eastern neighbor. Through a series of conflicts, Russian czars gained control over pieces of Baltic territories in the early eighteenth century, and officially annexed all Baltic lands into the Russian Empire in 1721. During that time the Baltic States were subject to Russification policies and remained under Russian control until the empire was overthrown in 1917.[2] After a brief period of independence from 1920 to 1940, the Baltic States saw their territories occupied by the Soviet army as part of a secret protocol in the Molotov-Ribbentrop Pact. They were briefly incorporated into Germany's Third Reich during the Second World War and became Soviet republics after 1945. It was not until the collapse of the Soviet Union in 1991 that the Baltic

---

[1] See, for example, Ted Galen Carpenter, "Are the Baltic States Next?" *National Interest*, March 24, 2014; and David A. Shlapak and Michael W. Johnson, *Reinforcing Deterrence on NATO's Eastern Flank: Wargaming the Defense of the Baltics*, Santa Monica, Calif.: RAND Corporation, RR-1253-A, 2016, p. 3. For a dissenting view that the Baltic States are at heightened risk of Russian aggression, see Robert Person, "6 Reasons Not to Worry About Russia Invading the Baltics," *Washington Post*, November 12, 2015.

[2] Russification efforts intensified under the reign of Czar Alexander III (1881–1894) and included the propagation of the Orthodox faith, imposition of the Russian language in schools and other official forums, and building of monuments to Russian leaders. See Peter Van Elsuwege, *Former Soviet Republics to EU Member States: A Legal and Political Assessment of the Baltic States' Accession to the EU*, Leiden: Brill, 2008, pp. 8–9; and Andres Kasekamp, *A History of the Baltic States*, Houndsmills, England: Palgrave Macmillan, 2010, pp. 84–87.

States regained their independence. Determined to break with their history of military vulnerability to Russia, the Baltic States quickly took steps to join NATO—a process that, after a decade of such efforts, resulted in their becoming members in 2004.[3]

The fall of the Soviet Union and accession to NATO did not mark the end of Russian influence in the Baltic States. Russia has leveraged the Baltic States' near total reliance on Russian oil to cut or threaten to cut supplies during times of tensions.[4] It has also established a "compatriot policy" meant to protect those ethnic Russians who found themselves outside Russia following the fall of the Soviet Union, including in Estonia and Latvia.[5] While there have been efforts to integrate these populations—for instance, Estonia cut its number of "noncitizens" by close to 60 percent between 1992 and 2000[6]—this issue still sparks occasional crises, and the Russian government has been suspected of using minorities as a lever to destabilize these states.

This chapter applies the deterrence framework outlined in Chapter Three to the specific case of U.S. and NATO efforts to deter a potential Russian aggression in the three Baltic States. It assesses Russia's motivation to undertake such an aggression; whether the United States and NATO have issued a clear and explicit message to Russia as to how they plan on responding to such an aggression, and what the costs would be for Russia; and, finally, how credible that message is for Russia. For each variable, we list the various indicators that can help establish whether a given variable is present at a low level (or absent); present at a medium level; or present at a high level. These indicators, however, do not imply a formal model. Rather, they give a sense of the evidence we used to make what is ultimately an analytical judgment.

Since the quality of a defender's deterrent posture relies heavily on the aggressor's intentions, as well as its perception of the defender's commitment to preventing an attack, this chapter relies on a number of Russian-language sources in addition to those in English, including official statements, interviews, comments to the press, and writings by high-level Russian officials and experts on strategic and military issues, as well as articles from several official or quasi-official strategic journals such as *National Strategy Issues* and *Russia in Global Affairs*.

---

[3] Ronald D. Asmus, *Opening NATO's Door: How the Alliance Remade Itself for a New Era*, New York: Columbia University Press, 2002, pp. 155–163, 228–238.

[4] Agnia Grigas, "Energy Policy: The Achilles Heel of the Baltic States," in Agnia Grigas, Andres Kasekamp, Kristina Maslauskaite, and Liva Zorgenfreija, *The Baltic States in the EU: Yesterday, Today and Tomorrow*, Studies and Reports No. 98, Paris: Notre Europe/Jacques Delors Institute, July 2013, pp. 70–71. In 1993, for instance, Russia cut gas supplies to Estonia to protest against a new residence law that affected ethnic Russians in Estonia; see Celestine Bohlen, "Russia Cuts Gas Supply to Estonia in a Protest," *New York Times*, June 26, 1993.

[5] Olga Oliker, Christopher S. Chivvis, Keith Crane, Olesya Tkacheva, and Scott Boston, *Russian Foreign Policy in Historical and Current Context: A Reassessment*, RAND Corporation Perspective, PE-144-A, 2015, p. 5.

[6] Marko Mihkelson, "Russia's Policy Toward Ukraine, Belarus, Moldova, and the Baltic States," in Janusz Bugajski, ed., *Toward an Understanding of Russia: New European Perspectives*, New York: Council on Foreign Relations, 2002, p. 107.

The chapter concludes with an assessment of U.S. and NATO deterrent posture toward Russia. Overall, we find that Russia's motivation to attack the Baltics (Variable 1) is relatively low, and the U.S. and NATO message to Russia warning it against an invasion of the Baltics is clear (Variable 2), but the credibility of the threat of military retaliation from the Alliance could be improved (Variable 3). Even with mixed credibility, however, the U.S. and NATO deterrent posture may be sufficient to hold back Russia given its limited ambitions in the Baltic States; and even if it deemed the risk low, Russia might be unwilling to take the chance of a NATO counterattack that might result in unbearable military costs.

How Motivated Is Russia?

Assuming that the Russian leadership is a reasonably unitary and rational actor, its decision to attack—or not—the Baltic States relies on an analysis of the costs and benefits to doing so. This section assesses what Russia would gain from aggression, examining whether it could create a more favorable strategic situation for Russia (Variable 1); whether the Russian leadership sees any urgency in changing the current strategic situation (Variable 2); the level of national interest that Russia sees the Baltic States as holding (Variable 3); and whether there is a sense in Moscow that action is urgently required or determined by a previous course of action (Variable 4).

Variable 1: The General Level of Dissatisfaction with the Status Quo and the Determination to Create a New Strategic Situation

Measurements: Public statements, national security strategies, military doctrines, recent behavior.

Russian leaders and analysts generally describe the "liberal" or U.S-led international order as threatening Russia's values, security, and interests.[7] This assessment comes after a series of developments, from NATO's enlargement to the development of ballistic missile defense, that have taken place over the past two decades; many among the Russian elite see these as evidence that Russian and U.S. interests are fundamentally divergent and feel that NATO consistently ignores legitimate Russian interests.[8] Accordingly, Russia has attempted to replace parts of the international system to make it more favorable to its interests—developing, for instance, the Collective Security Treaty Organization to balance NATO, and the Eurasian Economic Union as a counterpart to the European Union.

---

[7] Andrew Radin and Clinton Bruce Reach, *Russian Views of the International Order*, Santa Monica, Calif.: RAND Corporation, RR-1826-OSD, 2017, p. 32.

[8] For a good account of the gradual deterioration of U.S.-Russian relations from 1991 to 2014, see Stent, 2014. For an account of the evolution of NATO-Russia relations from a Russian perspective, see Sergei Oznobishchev, "Russia and NATO: From the Ukrainian Crisis to the Renewed Interaction," in Alexei Arbatov and Sergei Oznobishchev, eds., *Russia: Arms Control, Disarmament and International Security*, Moscow: IMEMO, 2016.

NATO's posture enhancement in Eastern Europe since 2014 has only exacerbated Russia's view of two colliding sets of interests. Commentary and public statements by Russian leaders and analysts describe NATO as using Ukraine as a pretext to return to Cold War–era containment policies.[9] Recent statements describe NATO-Russia relations as being at their lowest point since the end of the Cold War.[10]

Yet while Russia clearly seeks to revise what it perceives as an unfavorable strategic environment, there is little indication that it views aggressive actions in the Baltics as a promising means of achieving that purpose. Russia does not appear to see the strategic importance of the Baltics as on par with other former Soviet republics such as Georgia or Ukraine.[11] Russian officials have consistently denied hostile intentions toward the region, characterizing NATO's posture enhancements as responding to the "mythical" or "phantom" threats of Russian aggression; they have also dismissed NATO's concerns about Russia's incursions into states along its borders as baseless and pretextual. For instance, at the 2016 meeting of the Valdai Discussion Club, Putin described the "Russian military threat" to NATO and its allies as "unthinkable, foolish and completely unrealistic" in light of the size differential between Russia and Western powers combined, suggesting that it is a pretext to "pump new money into defense budgets at home, get allies to bend to a single superpower's interests, expand NATO and bring its infrastructure, military units and arms closer to our borders."[12] Additionally, there is little in the Russian public discourse to suggest that attacking the Baltic States would

---

[9] See, for example, Sergei Oznobischev, "Peretyagivanie Mira—chast' 1, Obschie Vyzovy I Ugrozy Vazhnee Krizisa na Ukraine" [Tug of peace—Part 1, common challenges and threats are more important than the Ukrainian crisis], *Voenno-Promyshlenny Kur'er*, July 23, 2014.

[10] "Vystupleniie Nachal'nika Genshtaba VS RF Generala Armii Valeriya Gerasimova na Konferentsii MCIS-2016" [Presentation of the chief of the general staff of the armed forces, Valery Gerasimov, at the MCIS-2016 Conference]," April 26, 2017; "Zayavlenie MID Rossii v Svyazi s Yubileinymi Datami v Otnosheniiakh Rossia-NATO" [Statement of the Ministry of Foreign Affairs in commemoration of key dates in Russia-NATO relations], May 26, 2017.

[11] Bryan Frederick, Matthew Povlock, Stephen Watts, Miranda Priebe, and Edward Geist, *Assessing Russian Reactions to U.S. and NATO Posture Enhancements*, Santa Monica, Calif.: RAND Corporation, RR-1879-AF, 2017, pp. 29–30; Radin and Reach, 2017, p. 11.

[12] "Vladimir Putin Meets with Members of the Valdai Discussion Club. Transcript of the Plenary Session of the 13th Annual Meeting," October 27, 2016. General Valery Gerasimov spoke at the MCIS-2016 Conference in similar terms, claiming that "to justify the policy of containment of Russia and to justify the demand for NATO, an old propagandistic trope—the Russian threat thesis—is being actively implanted in the collective consciousness of EU citizens"; "Vystupleniie Nachal'nika Genshtaba VS RF Generala Armii Valeriya Gerasimova na Konferentsii MCIS-2016," 2017. See also "Zayavlenie MID Rossii v Svyazi s Yubileinymi Datami v Otnosheniiakh Rossia-NATO," 2017. A number of Russian military and political analysts also suggest that the Baltic States instrumentally exaggerate their fear of Russia to secure U.S. and NATO support. On this theme, see, for example, Irina Batorshina, "Otnoshenii Pribaltiiskikh Respublik s Strategii Sderzhivaniia I Vovlecheniia Rossii (2014–2016)" [The attitude of the Baltic Republics to the strategy of containment and involvement of Russia (2014–2016)], *Problemy Natsional'noi Strategii*, Vol. 3, No. 42, 2017, and Prokhor Tebin, *A Tranquilizer with a Scent of Gunpowder: The Balance Between Russian and NATO Forces in Eastern Europe After 2014*, Valdai Papers No. 70, July 2017.

hold some strategic value. Bryan Frederick and colleagues, after reviewing extensively Russian strategic documents, conclude that "Any Russian decision to confront NATO militarily over the Baltics would not appear to come out of any existing vein of Russian strategic thinking."[13]

Russia's behavior in response to recent NATO actions in the Baltic region offers no clear indication of an aggressive intent.[14] As plans for the 2016 Warsaw Summit became known, Russia announced posture enhancements focused on its southwest region, along the Ukrainian border, with no comparable buildup in the vicinity of the Baltics.[15] While troops can be easily repositioned from one district to another,[16] such movements would take time and be visible, negating any surprise effect.[17] Moscow does maintain high-readiness units in the area and already possesses a significant local force advantage.[18] It has announced some plans for possible future enhancement of its posture in the area. But there is as yet no evidence of a continuing buildup to achieve even greater advantages.

While these various elements do not constitute evidence that Russia is *not* considering attacking the Baltic States, the fact that these countries are barely mentioned as a region of interest and the absence of clear concrete action to facilitate a potential invasion suggests that Russia's motivation to launch such military aggression is low.

Variable 2: The Degree of Fear That the Strategic Situation Is About to Turn Against Russia in Decisive Ways

Measurements: Measures of strategic balance, public statements, available intelligence on perceptions of strategic balance.

Despite Russia's efforts since 2009 to reform and modernize its military, the strategic balance between the United States and Russia still shows clear U.S. superiority. Russia spends ten times less than the United States on its national defense, has fewer military personnel, and has a lower count of key military equipment such as fighter

---

[13] Frederick et al., 2017, p. xiii. The authors add, "To be sure, Russia has taken and is continuing to take limited aggressive actions toward the Baltic States through political, media, intelligence, and cyber efforts. But we could identify no serious discussion of the strategic value of retaking part or all of the Baltic States, either for their intrinsic value or as a way of weakening NATO. This lack of discussion of the Baltics was in sharp contrast to other former Soviet states such as Ukraine and Georgia, which represent a much greater focus" (p. 77).

[14] Frederick et al., 2017, p. 55.

[15] Frederick et al., 2017, p. 55; Tebin, 2017, p. 13.

[16] Heather A. Conley, Kathleen H. Hicks, Lisa Sawyer Samp, Olga Oliker, John O'Grady, Jeffrey Rathke, Melissa Dalton, and Anthony Bell, *Evaluating Future U.S. Army Force Posture in Europe*, Washington, D.C.: Center for Strategic and International Studies, 2016, p. 3.

[17] Frederick et al., 2017, p. 10.

[18] See, for example, Michael Kofman, "Russian Military Buildup in the West: Fact Versus Fiction," *Russia Matters*, September 7, 2017.

aircrafts, ships, and submarines.[19] However, Russia's ability to provide for its most strategic missions—such as national defense, the maintenance of its nuclear power status, and interventions (when needed) in its near abroad—is high. Russia's intervention in Syria shows that it also has projection capabilities beyond its near abroad.

If Russian leadership became concerned that Russia might not be able to carry out one or more of these strategic missions in the near future, it could react by becoming increasingly willing to engage in risk-taking—with the goal, paradoxically, of preserving the status quo. Of note, Russia has opposed U.S. missile defense, which it sees as a potential threat to its second-strike capability. Russia's uneasiness with this program, however, dates to the 1990s,[20] and there are no recent developments in that area to suggest that Russia's view of the strategic balance has been fundamentally altered. Additionally, it is unclear whether Russia's fear of losing its nuclear power status would have any implications for the Baltic States specifically.

While public statements do indicate that Russia views NATO and U.S. actions in the region as evidence of aggressive intent and "provocations" aimed at drawing Russia into a confrontation,[21] it is unclear whether these views are mere political rhetoric or are sincerely held by Russian leaders.[22] There is also little indication that Russia fears imminent offensive action on the part of NATO, or that it feels compelled to preempt militarily some such action from NATO.[23] It is possible that the rising level and intensity of the rivalry between Russia and the West could affect these perceptions over time. The U.S. National Security Strategy has now identified Russia as a potential adversary and referred to a new era of great power strategic competition as the basic pattern in world politics.[24] To the extent that the competition worsens, and the United States and NATO begin taking more forceful actions to counter Russian strategic objectives, Moscow could come to view its situation with growing desperation. In particular, if NATO were to deploy assets capable of long-range strikes

---

[19] Andrew Tilghman and Oriana Pawlyk, "U.S. vs. Russia: What a War Would Look Like Between the World's Most Fearsome Militaries," *Military Times*, October 5, 2015.

[20] Stent, 2014, p. 30.

[21] See, for example, "Zasedanie Kollegii Federal'noi Sluzhby Bezobasnosti" [Meeting of the Collegium of the Federal Security Service], February 16, 2017. Putin remarked that "we are repeatedly provoked, so to speak, constantly provoked and being drawn into a confrontation. Attempts do not cease to meddle with our internal affairs with the goal of destabilizing socio-political situation in Russia itself." Russia's Representative to NATO has characterized NATO's actions in the region as "evidence of an approaching arms race," further noting, "Despite the assurances that these measures are not a provocation but a defensive reaction to changes in the area of security, the ongoing military development in NATO countries points in the opposite direction"; Alexander Grushko, "Speech by Russia's Permanent Representative to NATO Alexander Grushko at the Opening of the OSCE Annual Security Review Conference (ASRC) in Vienna, June 27, 2017."

[22] Frederick et al., 2017, p. 56.

[23] Frederick et al., 2017, p. 59.

[24] *National Security Strategy of the United States of America*, December 2017.

or assets that could degrade Moscow's anti-access/area denial capabilities, or if the United States developed or deployed new or additional technologies that posed an existential risk to Russian security, Russia's perception of its vulnerability would rise precipitously. For the moment, therefore, we judge this variable to be stable or positive, meaning that there is no indication that Russia sees itself as about to lose a key advantage or become particularly vulnerable and thus would direct an aggression at the Baltics to prevent this from happening. Yet emergent trends in world politics could change that situation.

Variable 3: The Level of National Interest Involved in the Specific Territory of Concern
Measurements: Historical relationship, public statements, available intelligence on perceived interests and intent.

Russian-Baltic relations have been poor since the collapse of the Soviet Union, with tensions focusing mainly on two issues: the treatment of Russian minorities by the governments of Estonia and Latvia; and conflicting views of history, with Russia defending the memory of the Red Army as a "liberator" of the Baltic States—a view the latter very much contests. Examples of tensions in recent years include the 2007 controversy surrounding the relocation of the Bronze Soldier of Tallinn; and the 2014 kidnapping of an Estonian intelligence officer, whom Russia accused of spying on its territory.[25] Since 2014, Russian aircraft have also conducted numerous violations of the Baltic States' airspace.[26]

Russian experts tend to view these problems as serious enough to preclude a normalization of relations in the foreseeable future.[27] Russia's view of the chance for opportunistic aggression against the Baltics, however, is constrained by the fact that they are solidly aligned with the West politically and militarily through their membership in the European Union and NATO. While Russian leadership has repeatedly mentioned NATO's expansion as a source of concern, and in theory Moscow would be delighted to take action that voids NATO control of the Baltics, there is no indication that Russia believes that such a reversal of NATO status is achievable in the Baltic States, 13 years after their accession to membership. Nor is their perceived security role as important as other countries: While Georgia and Ukraine could be part of a buffer zone between Russia and NATO members preventing what the Kremlin perceives as

---

[25] David M. Herszenhorn, "Russia and Estonia Differ over Detention," *New York Times*, September 5, 2014. The crisis was resolved a year later with the exchange of the Estonian officer for a former Estonian official jailed in Estonia on charges of spying for Russia; Jason Bush and David Mardiste, "Russia and Estonia Swap Alleged Spies," Reuters, September 26, 2015.

[26] In October 2016, Estonia accused Russia of having violated its airspace five times since the beginning of the year; "Russia Accused of Estonia Airspace Violations as Finland Signs Defense Pact With US," *Deutsche Welle*, October 8, 2016.

[27] Fyodor Lukyanov, "Stanet li Polsha modelyu dlya Baltii?" [Will Poland become the model for the Baltics?] *Russia in Global Affairs*, April 18, 2012; Batorshina, 2017.

a potential encirclement of Russia, the Baltic States can play no such role.[28] Overall, Frederick and colleagues note that "the Baltics hold relatively little intrinsic value for Russia" strategically and economically, and their importance is mostly symbolic[29]—which would explain why the most serious flare-up in their recent history has focused on the removal by Estonian authorities of the Bronze Soldier of Tallinn.[30]

In theory, an attack against the Baltics could have value for Russia as a test of NATO's commitment to collective defense. Yet there is no evidence that Russia is seriously considering such a high-risk move, which could entail devastating military and political costs. Besides, if the sole purpose of an attack was to test NATO, it would not have to take place in the Baltic States but could target instead another geographically close NATO ally such as Norway, Poland, or even Romania. Overall, the scenario of an attack on the Baltics to undermine NATO would only become plausible if NATO was already severely weakened, with internal tensions or a U.S. disinterest in the fate of the Alliance that would make Russia more likely to believe a gamble in the Baltics might succeed.[31]

Variable 4: The Urgent Sense of Desperation or Requirement to Act; Whether the Aggressor Is Locked into a Course of Action

Measurements: Public statements, available intelligence on perceived interests and intent, evidence of fears of general political instability or threat to the ruling regime.

While the Russian leadership is greatly concerned by the risk of internal instability,[32] as of mid-2017 Vladimir Putin still received high approval ratings and his position as political leader of the Russian Federation remained firmly established. A spring 2017 Pew Research Center survey reported that 78 percent of Russians expressed "some confidence" or "a lot of confidence" in Putin "to do the right thing regarding world affairs."[33] Support in Russia for Putin has declined since 2015 in a few specific areas, such as the handling of relations with the European Union, the Ukraine, and the

---

[28] Dmitry (Dima) Adamsky, "Cross-Domain Coercion: The Current Russian Art of Strategy," Proliferation Papers 54, Paris: Institut français de relations internationales, November 2015, p. 19.

[29] Frederick et al., 2017, p. 67.

[30] On this incident and its aftermath, see, for example, Martin Ehala, "The Bronze Soldier: Identity Threat and Maintenance in Estonia," *Journal of Baltic Studies*, Vol. 40, No. 1, March 2009, pp. 139–158; and Heather A. Conley, Theodore P. Gerber, Lucy Moore, and Mihaela David, *Russian Soft Power in the 21st Century: An Examination of Russian Compatriot Policy in Estonia*, Washington D.C.: Center for Strategic and International Studies, August 2011, pp. 4–8.

[31] Frederick et al., 2017, pp. 70–71, explore such a scenario, which they call "The West Weakened."

[32] See, for example, Dmitry Gorenburg, "Countering Color Revolutions: Russia's New Security Strategy and Its Implications for U.S. Policy," *Russian Military Reform*, September 15, 2014.

[33] Margaret Vice, "Russians Remain Confident in Putin's Global Leadership," Pew Research Center, June 20, 2017, p. 6.

United States, but approval ratings in all three areas remain high.[34] A Levada-Center poll conducted in March–April 2017 showed that 64 percent of respondents wanted Putin to remain president after the 2018 elections, while only 22 percent would have rather seen him replaced by someone else.[35] While Putin has had to weather public protests during his tenure as president—first in December 2011, then sporadically in 2012–2013, and most recently in March and June 2017—they have remained occasional outbursts that do not suggest that he could soon be unseated as leader.[36]

While it cannot be excluded in principle that, at some point in the future, the Russian leadership will find itself to be in a desperate situation and will decide to conduct a diversionary war in a hope to preserve its hold on power,[37] as of late 2017 there was no indication that this would happen soon—and if it did, it would likely be against an adversary less formidable than NATO. Finally, Russia does not appear to see itself locked in a course of action involving the Baltic States. While it may see itself as locked in a course of action regarding Ukraine—that is, an impossibility to back down from its current degree of commitment resulting in a continued involvement in a low-level war—this has no obvious or immediate impact on the Baltic States. Again, as suggested above, the intensifying rivalry between Russia and the United States carries the risk of altering this judgment, creating a context in which Moscow would come to believe that it had an urgent requirement to act in destabilizing and violent ways.

How Clear and Explicit Is the U.S. Deterrent Message?

Following Russia's incursions into Ukraine, both the United States and NATO have expressed their concerns about the security of the Baltic States and taken steps to dissuade Moscow from attempting a similar attack there. To be successful, the U.S. and NATO deterrent message must be clear and explicit: Russia should understand precisely what it is that the United States and NATO are seeking to prevent (Variable 1), and what actions they will take if Moscow attacks anyway (Variable 2). The message is more likely to be heard by Russia if it is expressed forcefully (Variable 3), and if it is repeated in the form of a warning whenever Russia takes threatening steps toward the Baltic States (Variable 4).

---

[34] The levels in 2017 were 73 percent, 67 percent, and 63 percent, respectively; Vice, 2017.

[35] Levada-Center, "Presidential Election," press release, May 29, 2017.

[36] Andrew S. Weiss, "Are Russian Protests a Threat to Putin?" KCRW radio broadcast, June 12, 2017.

[37] See Frederick et al., 2017, pp. 67–69, for an examination of such a scenario, which they title "Russia Lashes Out."

Variable 1: Precision in the Type of Aggression the United States Seeks to Prevent
Measurements: Statements by U.S. officials and in U.S. policy, availability of ambiguous forms of aggression to test clarity of limits.

While the United States and NATO have not explicitly detailed in their public statements and documents the types of Russian aggressive behavior they seek to deter in the Baltic States, this information is conveyed indirectly through their condemnation of the territorial violations of Ukraine; the mention of previous instances of aggression such as the Soviet occupation of the Baltic States; and the affirmation of their commitment to collective defense (which would be triggered by an "armed attack" against a NATO member). For instance, in his remarks in Estonia in 2014, President Barack Obama stated,

> [Russia's aggression against Ukraine] is a brazen assault on the territorial integrity of Ukraine, a sovereign and independent European nation. . . . Countries like Estonia, Latvia, and Lithuania are not "post-Soviet territory." You are sovereign and independent nations with the right to make your own decisions. No other nation gets to veto your security decisions. . . . An attack on one is an attack on all. . . . You lost your independence once before. With NATO, you will never lose it again.[38]

On various occasions, U.S. and NATO officials have implied that any violation of territorial sovereignty of a NATO member would not be tolerated and would trigger Article V of the Washington Treaty.[39] While Article V does not indicate exactly what constitutes an "armed attack,"[40] NATO officials have attempted to clarify this point. Shortly after Russia's invasion of Crimea, SACEUR GEN Philip Breedlove stressed the need for NATO to define how the Alliance views and responds to nontraditional forms of warfare. NATO, he said, needs to

> mature the way we think about cyber, the way we think about irregular warfare, so that we can define in NATO what takes it over that limit by which we now have to react. . . . It is illustrative for us to look at this form of warfare we're seeing from Russia and how we will react to it in the future.[41]

---

[38] Barack Obama, "Remarks at Nordea Concert Hall in Tallinn, Estonia," September 3, 2014b.

[39] Reuters staff, "Merkel Pledges NATO Will Defend Baltic Member States," Reuters, August 18, 2014; Obama, 2014b; Barack Obama, "Remarks by President Obama and Leaders of Baltic States in Multilateral Meeting," September 3, 2014c; Robin Emmott and Sabine Siebold, "NATO Agrees to Reinforce Eastern Poland, Baltic States Against Russia," Reuters, July 7, 2016; Joe Biden, "Remarks by Vice President Joe Biden at the National Library of Latvia," August 24, 2016; North Atlantic Treaty Organization, "Press Conference by NATO Secretary General Jens Stoltenberg Ahead of the Meeting of NATO Heads of State and Government," May 24, 2017c; Joseph Biden, "Remarks by the Vice President to Enhanced Forward Presence and Estonian Troops," July 31, 2017.

[40] North Atlantic Treaty Organization, The North Atlantic Treaty, April 4, 1949.

[41] John Vandiver, "Breedlove: NATO Must Redefine Responses to Unconventional Threats," *Stars and Stripes*, July 31, 2014.

NATO member states have since decided that certain "serious cyber attacks" can "trigger the Article V collective defense clause," but that "cyber is not something that always triggers Article V."[42] NATO has not indicated what qualifies as a "serious cyber attack."[43] Regarding the second issue in Breedlove's comments—irregular warfare—the Alliance has since identified "hybrid warfare" as a "security challenge" and has agreed that while the "primary responsibility to respond to hybrid attacks rests with the targeted nation . . . [the] Council could decide to invoke Article V of the Washington Treaty" in the event of a hybrid, or gray zone, attack.[44] In essence, the Alliance recognizes that not all attacks take the form of a conventional military incursion onto a member's territory; and that while they do not exclude invoking collective defense against cyber or unconventional attacks, this response will not be automatic. These statements signal to Russia that unconventional attacks against a NATO member could provoke a military response, though it remains unclear which ones would meet this threshold. It also means that there is limited utility for Moscow to use such attacks to test NATO, since NATO's commitment is not automatically engaged.

Variable 2: Clarity in the Actions That Will Be Taken in the Event of Aggression
Measurements: Public statements by U.S. officials, statements in U.S. strategy documents, policies and actions to support clarity (e.g., troop deployments).

U.S. officials have expressed their commitment to collective defense during their frequent visits to the Baltic States.[45] For instance, Vice President Joseph Biden addressed the presidents of Latvia and Lithuania in Vilnius in March 2014, shortly after the invasion of Crimea, and stated,

> The reason I traveled to the Baltics was to reaffirm our mutual commitment to collective defense. President Obama wanted me to come personally to make it clear what you already know, that under Article V of the NATO treaty, we will respond. We will respond to any aggression against a NATO ally.[46]

---

[42] North Atlantic Treaty Organization, "Press Conference by NATO Secretary General Jens Stoltenberg Following the North Atlantic Council Meeting at the Level of NATO Defence Ministers," June 14, 2016a; North Atlantic Treaty Organization, "Remarks by NATO Secretary General Jens Stoltenberg at the Elliott School of International Affairs, George Washington University," April 13, 2017b.

[43] North Atlantic Treaty Organization, 2017b.

[44] North Atlantic Treaty Organization, "Warsaw Summit Communiqué," July 9, 2016b. NATO ministers had already adopted a new hybrid warfare strategy in December 2015 to "prepare, deter, and defend against hybrid threats," and pledged in 2016 and 2017 to work closely with the European Union to counter these threats.

[45] See Joseph Biden, Dahlia Grybauskaite, and Andris Berzins, "Remarks to the Press by Vice President Joe Biden, President Dahlia Grybauskaite of Lithuania, and President Andris Berzins of Latvia," March 19, 2014; Obama, 2014b; Obama, 2014c; Biden, 2016; and Biden, 2017.

[46] Biden, Grybauskaite, and Berzins, 2014.

Yet there is no specific description in the Washington Treaty of what such defense would entail. Article V states that NATO will respond "with action as it deems necessary, including the use of armed force, to restore and maintain the security of the North Atlantic area."[47] In other words, the military response is only one of many options, and statements by U.S. officials have generally remained vague when it comes to describing what the United States would do ("respond,"[48] "step up,"[49] or "hold Russia accountable for its actions"[50]) if its allies came under attack, although President Obama mentioned on at least one occasion that "the NATO alliance, including the Armed Forces of the United States of America," would come to Estonia's aid.[51] U.S. officials might want to keep a range of options open rather than commit publicly to a military response; they may also be reluctant to commit to a decision that will be impacted, if and when the time comes, by the decisions that other allies will make.

While the United States is not clearly threatening military retaliation, in practice it has taken steps to signal its willingness to respond to a potential Russian incursion in the Baltics by military means. In June 2014 the United States announced a $1 billion European Reassurance Initiative, or ERI (later renamed the European Deterrence Initiative), to fund an increased presence of U.S. troops, pre-positioning of equipment, more robust training and exercises, and improvements to facilities and infrastructure in Europe.[52] In the speech that announced this initiative, President Obama made clear that NATO's military deployments in Eastern Europe were aimed at giving some "teeth" to the U.S. commitment to Article V:

> Article V is clear—an attack on one is an attack on all. And as allies, we have a solemn duty—a binding treaty obligation—to defend your territorial integrity. And we will. . . . Poland will never stand alone. (Applause.) But not just Poland—Estonia will never stand alone. Latvia will never stand alone. Lithuania will never stand alone. Romania will never stand alone. (Applause.) These are not just words. They're unbreakable commitments backed by the strongest alliance in the world and the armed forces of the United States of America—the most powerful military in history. (Applause.) You see our commitment today. In NATO aircraft in the skies of the Baltics. In allied ships patrolling the Black Sea. In the stepped-up exer-

---

[47] North Atlantic Treaty Organization, The North Atlantic Treaty, Article V.

[48] Biden, Grybauskaite, and Berzins, 2014.

[49] Biden, 2016.

[50] Biden, 2017.

[51] Obama, 2014b.

[52] The White House, Office of the Press Secretary, "Fact Sheet: European Reassurance Initiative and Other U.S. Efforts in Support of NATO Allies and Partners," June 3, 2014.

cises where our forces train together. And in our increased and enduring American presence here on Polish soil.⁵³

In a July 2017 speech, Vice President Mike Pence also hinted at potential military action on the part of the United States by mentioning as evidence of the U.S. commitment to the Alliance the following:

> Under the leadership of President Trump, the United States will make the strongest fighting force in the history of the world even stronger. . . . The President has already signed the largest increase in military spending in nearly a decade. And we've called on Congress to pass one of the largest investments in defense spending since the days of the Cold War.⁵⁴

During its 2014 Wales Summit, NATO similarly adopted a Readiness Action Plan to enhance the security of the Baltic States. The plan was described as a "package of necessary measures to respond to . . . the challenges posed by Russia and their strategic implications,"⁵⁵ and included a threefold increase in the size of the NATO Response Force, as well as the creation of a Very High Readiness Joint Task Force deployable on short notice. Since then, NATO members have taken further steps to demonstrate their willingness to deploy military assets in the defense of the Baltics, including four multinational battalion-size battle groups. Following Russia's invasion of Ukraine, NATO has launched new exercises in the Baltic region such as Steadfast Javelin and increased the size and scope of existing exercises like Baltic Operations and Saber Strike. For instance, the U.S.-led Saber Strike was first conducted in 2011 with 2,000 personnel, but its 2017 iteration saw the participation of over 11,000 personnel.⁵⁶ The 2017 Saber Strike exercise also involved for the first time a defense of the Suwalki Gap on the border between Lithuania and Poland—an area that, if seized by Russia, would isolate the Baltic States from their NATO allies.⁵⁷ Rehearsing such scenarios sends a signal to Russia that the United States and NATO are preparing against all contingencies and suggests they intend to defend this area if it is under attack.

---

⁵³ Barack Obama, "Remarks of President Obama at 25th Anniversary of Freedom Day," June 4, 2014a.

⁵⁴ Biden, 2017.

⁵⁵ North Atlantic Treaty Organization, "Wales Summit Declaration," September 5, 2014.

⁵⁶ U.S. Army Europe Public Affairs, "Exercise Saber Strike 2012 Demonstrates International Cooperation in Action," March 15, 2012; Brooks Fletcher, "Saber Strike 2013 a Demonstration of Multinational Partnership in the Baltics," U.S. Army Europe Public Affairs, June 13, 2013; U.S. Army Europe, "Exercise Saber Strike 14 Demonstrates International Cooperation," June 10, 2014; U.S. Army Europe Public Affairs, "Exercise Saber Strike 15 Demonstrates International Cooperation Capabilities," June 1, 2015; U.S. Marines, Combat Logistics Regiment 2, "U.S. NATO Allies Conduct Large-Scale Exercise to Defend Baltics," June 20, 2016; Tryphena Mayhugh, "U.S., NATO Conclude Saber Strike 17 Exercise," U.S. Department of Defense, June 26, 2017.

⁵⁷ Andrius Sytas, "NATO War Game Defends Baltic Weak Spot for First Time," Reuters, June 18, 2017.

Variable 3: Forceful Communication of These Messages to Outside Audiences, Especially Potential Aggressor(s)

Measurements: Number and profile of statements, direct communication to the aggressor in diplomatic forums, evidence that the aggressor has heard and understood the commitment.

The clear warning to Russia that an attack on the Baltic States would trigger Article V was made publicly on repeated occasions at the highest levels of the U.S. administration.[58] While these statements were meant to assure the Baltic States of U.S. support, they also aimed at communicating to Russia that the United States is committed to defending its allies, as the following speech delivered by Vice President Biden in Latvia in 2016 makes clear:

> Aggression still happens in Europe and we must be ready to answer that aggression. That's what Article V means—we are all prepared to step up. Not just the United States and the Baltic States, but all of us. An attack on one is an attack on all. Period. End of sentence. It's that basic, it's that simple. And we want you to know, *we want Moscow to know*, that we mean what we say.[59]

Similarly, in the communiqué issued at the 2016 Warsaw Summit, NATO heads of state noted that their purpose in establishing the enhanced forward presence in the Baltic States and Poland was to "unambiguously demonstrate, as part of our overall posture, Allies' solidarity, determination, and ability to act by triggering an immediate Allied response to any aggression."[60] The document does not explicitly mention Russia as a threat, but it points to the "changed and evolving security environment" as the impetus for the Alliance's enhancements.[61]

While it is difficult to confirm whether the Russian leadership has understood the U.S. and NATO commitment, there is little doubt that it is closely attuned to U.S. and NATO statements that relate to Russia and their intentions toward Russia. Additionally, the existence of high-level contacts between Russian and NATO military leaders increases the likelihood that Russia hears the NATO messages—although misunderstandings can never be ruled out.[62]

Some early statements from President Donald Trump appeared to call into question these messages of unquestioned commitment, communicating instead an intention to demand better burden sharing from key allies as the price of contin-

---

[58] See, for example, Obama, 2014c; and Biden, 2017.

[59] Biden, 2016, emphasis added.

[60] North Atlantic Treaty Organization, 2016b.

[61] North Atlantic Treaty Organization, 2016b.

[62] See, for example, "Nachal'nik General'nogo shtaba Vooruzhennyx Sil RF general armii Valerii Gerasimov provel telefonny razgovor s predsedatelem Voennogo Komiteta NATO generalom Petrom Pavelom" [Chief of the general staff of the armed forces Valerii Gerasimov conducted a telephone conversation with chairman of the NATO Military Committee Petr Pavel], March 3, 2017.

ued U.S. Article V guarantees.[63] Many other senior members of the administration qualified these comments and broadcast a continued U.S. intention to abide by the commitments of the NATO Alliance, and President Trump himself later publicly endorsed such an intention.[64] To the extent that the current administration's hesitations reflect longer-term trends in U.S. politics—including a growing irritation with foreign commitments—the message broadcast by the United States is likely to become more conditional and ambiguous. Trends in U.S. politics, therefore, suggest that while this variable currently falls into the "healthy deterrence" category, over time it may become more questionable.

Variable 4: Timely Response to Warning with Clarification of Interests and Threats
Measurements: Quality of warning system, willingness of political leaders to use warning for deterrent effect.

Russia's aggression toward Ukraine generally elicited a reaction of surprise, including in the United States.[65] Russia's unexpected move underlined how much uncertainty exists regarding Putin's intentions and how far he might be willing to go to upset and revise the existing international order. Moscow's growing military involvement in the war in Syria has strengthened this uncertainty. Its intervention in Syria has showed a willingness—largely unprecedented in the history of Russia since the fall of the Soviet Union—to play a direct military role in a conflict outside Russia's immediate neighborhood. The result of Putin's actions is that Russia is now, more than ever, under close scrutiny. In particular, U.S. and European observers are paying close attention to military exercises because Moscow has used them in the past as a cover for aggressive military action.[66] Since 2014 such exercises have also been larger in scope and more frequent.[67]

In the wake of the Ukraine crisis, the United States and NATO adopted a series of reassurance measures toward the Alliance's Central and Eastern European members.

---

[63] Max Fisher, "Donald Trump's Ambivalence on the Baltics Is More Important than It Seems," *New York Times*, July 21, 2016; Justin McCurry, "Trump Says U.S. May Not Automatically Defend NATO Allies Under Attack," *Guardian*, July 21, 2016.

[64] Jeremy Herb, "Trump Commits to NATO's Article 5," *CNN*, June 9, 2017.

[65] See, for example, Magnus Christiansson, "Strategic Surprise in the Ukraine Crisis: Agendas, Expectations, and Organizational Dynamics in the EU Eastern Partnership Until the Annexation of Crimea 2014," master's thesis, Swedish National Defence College, August 2014; Kristin Ven Bruusgaard, "Crimea and Russia's Strategic Overhaul," *Parameters*, Fall 2014, p. 84; Bob Work, "The Third U.S. Offset Strategy and Its Implications for Partners and Allies," U.S. Department of Defense, January 28, 2015; and Heather A. Conley, "Russia's Influence on Europe," in Craig Cohen and Josiane Gable, eds., *2015 Global Forecast: Crisis and Opportunity*, Washington, D.C.: Center for Strategic and International Studies, 2014, p. 28.

[66] For the case of Crimea in 2014, see, for example, Michael Kofman, Katya Migacheva, Brian Nichiporuk, Andrew Radin, Olesya Tkacheva, and Jenny Oberholtzer, *Lessons from Russia's Operations in Crimea and Eastern Ukraine*, Santa Monica, Calif.: RAND Corporation, RR-1498-A, 2017.

[67] Conley et al., 2016, p. 3.

These measures also communicated to Russia that a similar behavior against a NATO ally—including the Baltic States—would not be tolerated. Subsequently, the United States specifically condemned incidents that occurred in the Baltic region, such as Russian planes coming too close to U.S. ships and aircraft, as well as violations of the Baltic States' airspace. However, the United States does not appear to have used these opportunities to reiterate its deterrent message, likely because these incidents were considered mere Russian provocations rather than the precursors of an attack against the Baltic States. Accordingly, it made sense for the United States to downplay these incidents rather than escalate, since miscalculation, accidents, and escalation are precisely what Washington seeks to avoid in these situations.[68]

Is the U.S. Deterrent Message Credible and Convincing?

Even if a deterrent message is expressed clearly by the defender and similarly received by the aggressor, the overall deterrence posture of the defender might be weak if the potential aggressor does not think that the defender will deliver on its promise. Some of the factors that damage credibility are the same as those that that damage clarity—for instance, contradictory statements and mixed messages blur the defender's intent and suggest to the aggressor that there is no consensus behind the threat being communicated. In this section we review four key elements that make a deterrent message credible: the actual and perceived strength of the military capabilities present in the Baltic region (Variable 1); the degree of automaticity of a U.S.-NATO response (Variable 2); the perceived strength of the political commitment, on the part of the defender, to fulfill the deterrent threat (Variable 3); and the degree of national interests engaged by the United States (and, to a lesser extent, other NATO members) in the Baltic States (Variable 4).

Variable 1: The Actual and Perceived Strength of the Local Military Capability to Deny the Presumed Objectives of the Aggression

Measurements: Objective strength of capability, evidence of potential aggressor perception of capability, type and character of local forces, evidence of close integration with a local partner and concepts of operation to deny quick victory, quality of local and regional basing infrastructure to support operations.

The Baltic States' military capabilities have improved since they became NATO members, and these countries stand at a high level of readiness.[69] They have developed

---

[68] Lisa Ferdinando, "Russian Airspace Violations in Nordic-Baltic Regions Dangerous, Work Says," U.S. Department of Defense, October 6, 2016; U.S. European Command, "Navy Ship Encounters Aggressive Russian Aircraft in Baltic Sea," U.S. Department of Defense, April 13, 2016a.

[69] Duncan Long, Terrence Kelly, and David C. Gompert, eds., *Smarter Power, Stronger Partners*: Vol. II, *Trends in Force Projection Against Potential Adversaries*, Santa Monica, Calif.: RAND Corporation, RR-1359/1-A, 2017, p. 136.

"total defense" strategies that involve not just their military forces but also the mobilization of their entire societies to resist aggression and deny the attacker an easy victory.[70] Yet Russia's capabilities dwarf those of all three countries, and the Baltic States' force structure since 2004 has focused more on supporting NATO's out-of-area operations than on building up territorial defense.[71]

NATO's reassurance efforts since 2014, such as the deployment of the four multinational battle groups, have aimed at reinforcing the Baltic States' side in a balance of power that is largely tipped toward Russia. Since September 2015, the activation of NATO Force Integration Units in each of the three Baltic States has further enhanced the Alliance's ability to deploy forces in these countries if needed, and has improved integration between the national forces of the Baltic States and those of their NATO allies.[72]

Several studies, however, underline the limits of what such reinforcements could achieve. David Johnson and Michael Shlapak conclude, based on more than 20 war games involving Russian aggression in the Baltics, that NATO would have a hard time denying a quick victory to Russia were Russia determined to attack.[73] Several other studies support the similar conclusion that adequate defense of the Baltics would require substantial investments and deployments beyond the four battle groups, although they differ on the precise mix of forces that might provide a convincing deterrent.[74] To come to the military support of the Baltics, NATO forces would have to overcome various obstacles ranging from the geographic isolation of the Baltics and the risk of bypassing Kaliningrad to the lack of logistical lines across Europe.[75] Examining a potential war scenario between Russia and Estonia, Duncan Long, Terrence Kelly, and David Gompert note,

> Once a conflict seems imminent, any deployment of forces directly to the Baltics by air or sea is at great risk. Should deterrence fail and should significant NATO forces not be on the ground to contest a Russian invasion, NATO will be confronted with rolling back forces already enveloped in an A2AD umbrella. It will

---

[70] Jan Osburg, Stephen J. Flanagan, and Marta Kepe, "How to Deter NATO's Greatest Fear: A Russian Invasion of the Baltic States," *National Interest*, November 22, 2016.

[71] Eoin Micheál McNamara, "Securing the Nordic-Baltic Region," *NATO Review Magazine*, n.d.

[72] North Atlantic Treaty Organization, "NATO Force Integration Units," fact sheet, September 2015.

[73] Shlapak and Johnson, 2016, pp. 4–6.

[74] See Conley et al., 2016; Franklin D. Kramer and Bantz J. Craddock, *Effective Defense of the Baltics*, Washington D.C.: Atlantic Council, 2016; and Anderson R. Reed, Patrick J. Ellis, Antonio M. Paz, Kyle A. Reed, Lendy Reenegar, and John T. Vaughn, *Strategic Landpower and a Resurgent Russia: An Operational Approach to Deterrence*, Carlisle, Penn.: U.S. Army War College Strategic Studies Institute, 2016. For a comprehensive comparison of the force posture recommendations of these four studies, as well as Shlapak and Johnson's, see Frederick et al., 2017, pp. 17–22.

[75] Shlapak and Johnson, 2016, pp. 3–4.

take a long time for the United States to move forward sufficient heavy forces to isolate Kaliningrad and march through the Baltics along a broad front of Russian territory."[76]

The perception from Moscow appears to follow somewhat similar lines. NATO's posture enhancement in Eastern Europe has been described as threatening Russia's security, increasing the potential for conflict in the region, spurring a wasteful arms race, and violating NATO and Russia's Founding Act.[77] The Russian leadership has taken issue with the "rotational" NATO presence in the Baltics, arguing that uninterrupted rotation is indistinguishable from permanent stations of military capabilities and violates the Founding Act's pledge against the permanent stationing of additional substantial combat forces.[78] Yet the Russian leadership and analysts do not appear to see the NATO posture enhancements announced at the Warsaw summit as fundamentally changing the local balance of power.[79]

The status of deterrence in the Baltics could also be affected by Russian concepts governing the role of nuclear weapons—both as instruments of coercion and as potential war fighting tools. Russia pursues a concept of "strategic deterrence" involving both nonmilitary tools (such as cyber and disinformation) and nuclear coercive threats to gain advantage over potential adversaries.[80] Some analysts suggest that Russia has developed a detailed concept of "escalate to de-escalate," in which it threatens, or actually employs, low-yield nuclear weapons to achieve decisive operational effects and shock NATO into withdrawal.[81] Russia has certainly been investing in new, smaller-scale nuclear weapons in recent years, and its recent military doctrines place significant emphasis on the deterrent role of its nuclear arsenal.

Yet its doctrine also tends to treat nuclear weapons as ultimate forms of deterrence largely reserved for protection of the homeland in the event of existential threats. Suggestions of the role of nuclear weapons in regional conflict emphasize coercive value; there is little evidence that Moscow is prepared to use nuclear weapons early as part of

---

[76] Long, Kelly, and Gompert, 2017, p. 161.

[77] "Vystupleniie Nachal'nika Genshtaba VS RF Generala Armii Valeriya Gerasimova na Konferentsii MCIS-2016," 2017.

[78] See, for example, "Zayavlenie MID Rossii v Svyazi s Yubileinymi Datami v Otnosheniiakh Rossia-NATO," 2017.

[79] Frederick et al. 2017, p. 57; Tebin, 2017, p. 10, emphasizes the low military value of the NATO battalions deployed in Poland and the Baltic States, underlining instead their political value in communicating a message of resolve to Russia.

[80] Kristin Ven Bruusgaard, "Russian Strategic Deterrence," *Survival*, Vol. 58, No. 4, November–December 2014, pp. 7–26.

[81] Mark B. Schneider, "Escalate to De-Escalate," *Proceedings*, February 2017; Elbridge Colby, "Russia's Evolving Nuclear Doctrine and Its Implications," Foundation for Strategic Research, January 2016.

a strategy of opportunistic offensive warfare. Suggestions that Russia may be prepared to escalate to low-level nuclear use to force NATO de-escalation may be exaggerated.[82]

Russian nuclear threats could affect the operational outcome by ruling out specific categories of NATO response. Moscow could imply or directly state, for example, that significant precision strikes against Russian territory, including Kaliningrad, might trigger a highly escalatory response. NATO would then be confronted with a difficult choice: provide Russian forces with what amounts to a sanctuary close to the combat area, or risk nuclear escalation. Such threats could affect the local military outcome.

Variable 2: The Degree of Automaticity of the U.S.-NATO Response, Including Escalation to a Larger Conflict.
Measurements: Alliance commitments and public pledges, significance of local forces, likely scale of loss of life in the local fight, integration into regional commands that ensure rapid escalation, political conditions/willpower in the United States, factors likely to make accidental escalation likely or unlikely.

According to Article V of the Washington Treaty, an "armed attack against one or more" allies should automatically lead NATO to "assist the Party or Parties so attacked by taking forthwith, individually and in concert with the other Parties, such action as it deems necessary, including the use of armed force."[83] As discussed above, however, the definition of "armed attack" is unclear. Additionally, if Russia were to attack, it might try to obscure its involvement, as it did initially in the east of Ukraine, potentially delaying a NATO reaction until responsibilities were clearly established. Finally, the treaty leaves open the type of response that allies can choose to assist the attacked party. In other words, the treaty makes clear that a military response to a Russian attack against the Baltic States would not be automatic—a precaution the United States took when it drafted the treaty originally, to ensure that it would not be dragged against its will into another European war.[84] More generally, the fact that a military response to an armed attack is likely but not automatic gives the United States and each NATO member some leeway to decide what is the most appropriate response. Yet the presence of battalions on the ground acting as trip wire in the event of a Russian attack would make conflict inevitable. Under these circumstances, therefore, there would be an automatic response, and an escalation of the conflict, on the part of NATO.

Under the administrations of George W. Bush and Barack Obama, Russia perceived NATO's commitment to collective defense in Eastern Europe to be high: Russia must, to some extent, have believed that NATO might use military power against

---

[82] Olga Oliker, *Russia's Nuclear Doctrine: What We Know, What We Don't, and What That Means*, Center for Strategic and International Studies, May 2016; Kristin Ven Bruusgaard, "The Myth of Russia's Lowered Nuclear Threshold," *War on the Rocks*, September 22, 2017.

[83] North Atlantic Treaty Organization, The North Atlantic Treaty, April 4, 1949.

[84] North Atlantic Treaty Organization, "Collective Defense—Article 5," March 22, 2017a.

Russia; otherwise, its opposition to NATO membership for Georgia or Ukraine is harder to explain.[85] It is unclear to what extent the change in U.S. administration has modified this perception, if at all. As candidate, Donald Trump publicly made a U.S. intervention to defend a NATO ally (which had been described by the interviewer as "Estonia or Latvia, Lithuania") conditional on their "paying their bills."[86] He famously described NATO as "obsolete," a position he later reversed as President.[87] And his remarks at the unveiling of the Article V and Berlin Wall memorials in May 2017 did not include any clear affirmation of the U.S. commitment to collective defense.[88] Two months later, however, Vice President Pence stated, on his visit to the Baltic States,

> Be assured: The United States rejects any attempts to use force, threats, intimidation, or malign influence in the Baltic States or against any of our treaty allies—and under President Donald Trump, the United States of America will stand firmly behind our Article V pledge of mutual defense—and the presence of the U.S. Armed Forces here today proves it.[89]

In June 2017 President Trump similarly expressed support for the Alliance, stating, "Yes, absolutely, I'd be committed to Article V."[90] These mixed messages on the sanctity—or conditionality—of the U.S. commitment to NATO are compounded by a U.S. policy toward Russia that ranges from diplomatic overtures to Russia being described as a "grave security concern" that NATO should focus on in the future.[91]

From the Russian side, officials' statements have welcomed Trump's rhetoric promising less adversarial relations with Russia but eschewed excessive optimism.[92] Putin himself offered the view that NATO remains an instrument of U.S. foreign policy.[93]

---

[85] Frederick et al. 2017, pp. 31, 62.

[86] "Transcript: Donald Trump on NATO, Turkey's Coup Attempt and the World," *New York Times*, July 21, 2016. On Russia's hope that President Trump would be a more "transactional" leader than his predecessor, see Marek Menkiszak, *Russia's Best Enemy: Russian Policy Towards the United States in Putin's Era*, Point of View No. 62, Warsaw: Center for Eastern Studies, February 2017, p. 46.

[87] Peter Baker, "Trump's Previous View of NATO Is Now Obsolete," *New York Times*, April 13, 2017.

[88] Trump, Donald J. "Remarks by President Trump at NATO Unveiling of the Article 5 and Berlin Wall Memorials—Brussels, Belgium," May 25, 2017.

[89] Biden, 2017.

[90] The White House, Office of the Press Secretary, "Remarks by President Trump and President Iohannis of Romania in a Joint Press Conference," June 9, 2017.

[91] Trump, 2017.

[92] See for instance, "Peskov rasskazal pro ozhidaniia ot administratsii Trampa" [Peskov addressed expectations of the Trump administration], *Rossiiskaia Gazeta*, December 21, 2016. Vladimir Putin's Press Secretary Dmitry Peskov commented that "we do not expect to solve all problems, we do not expect that America will imminently repudiate NATO enlargement or the situation of NATO military infrastructure near our borders."

[93] "St. Petersburg International Economic Forum Plenary Meeting," June 2, 2017.

Russian officials including Putin also cited Trump's call for increasing the financial contribution of other NATO members as evidence of the Alliance's aggressive intent.[94] They also described President Trump's actions as constrained by Congress and the Democratic Party to follow "traditional American policy," suggesting a continuity between the Obama and Trump administrations.[95] Moreover, insofar as Russian officials note a split within NATO, the United States is often portrayed as the more aggressive member of the Alliance, pushing other European powers—even those who might be more accommodating to Russia, in part owing to economic interests—toward greater conflict with Russia. The new economic sanctions against Russia signed into law by President Trump on August 2, 2017, appear to have crushed whatever hopes the Russian elite may have had left for better U.S.-Russia relations under the new U.S. administration.[96]

The role of Russian nuclear weapons could affect the automaticity and degree of a U.S. or NATO response to aggression. Were Moscow to intervene in the Baltics while making statements implying the ready usability of nuclear weapons, it could affect the rapidity of NATO responses as the Alliance struggled to deal with the escalatory risks involved. Yet unless Russian actions caused NATO to fundamentally change its doctrine and military plans before a conflict, such nuclear posturing would not appear to undercut the expectation of essentially automatic NATO conventional military responses under Article V: U.S. and NATO aircraft and maritime assets, for example, would presumably be engaged in the fight from the first hours. Russian nuclear coercion would have the difficult job of convincing a military alliance already drawn into a wide-ranging fight to back off; such withdrawal under fire would be politically

---

[94] For instance, Putin remarked: "The United States is demanding that its allies increase their military spending, while maintaining at the same time that NATO has no plans to attack anyone. If you are not going to attack anyone, why increase military spending? Of course, this raises additional questions on our part"; "St. Petersburg International Economic Forum Plenary Meeting," 2017. In an interview with *Le Figaro*, Putin again displayed skepticism about NATO's interest in improving relations with Russia, when he acknowledged discussions of such improvement at the NATO summit and asked, "Then why are they increasing their military spending? Whom are they planning to fight against?" "Inverviu Vladimira Putina frantsuzkoi gazete Le Figaro" [Vladimir Putin's interview with the French newspaper Le Figaro], May 31, 2017. Likewise, Russia's permanent representative to NATO stated, "NATO appears to be planning to abandon the ... dividends [of peace afforded by the end of the Cold War] and to return to the past. Otherwise, why is it not satisfied with Europe's current defence spending, which has reached 250 billion euros?" "Speech by Russia's Permanent Representative to NATO," 2017.

[95] "Inverviu Vladimira Putina frantsuzkoi gazete Le Figaro," 2017. Putin observed that notwithstanding Trump's words about better relations with Russia, Russians understand that policy is driven by "people who lost the election, but do not want to reconcile themselves to this, and unfortunately, use the anti-Russian card" for their own political purposes. After Trump signed the bill imposing sanctions on Russia, Prime Minister Dmitry Medvedev noted on his Facebook page that "the Trump administration demonstrated total powerlessness, abdicating its executive power to Congress in the most demeaning fashion." See also "Mir bez Illyuziy i Mifov" [Peace without illusions or myths], *Rossiiskaya Gazeta*, January 15, 2017; and "Rossiyane stali khuzhe otnosit'sya k Trampu" [Russians' opinion of Trump worsened], *Rossiiskaya Gazeta*, August 4, 2017.

[96] For instance, according to Prime Minister Medvedev, Trump's signing the sanctions bill spelled the "end to the hopes for improving our relations with the new U.S. administration." "Rossiyane stali khuzhe otnosit'sya k Trampu," 2017.

costly for many European countries, as well as being operationally challenging, if not impossible.

At the same time, the escalatory risks implied by aggressive Russian nuclear posturing cut both ways. As much as it could cause NATO to pause before putting into effect certain military responses, lowering the nuclear threshold would also risk making a Baltic adventure much costlier for Russia if the coercive gambit failed. Most narratives that outline a possible Russian invasion imagine scenarios in which Moscow convinces itself it can get away with a fait accompli without sparking a larger war. Threatening early use of nuclear weapons in response to NATO conventional operations would undermine such a concept, raising the potential of a local or more general nuclear conflict. Unless Russia became convinced that NATO would simply give up and depart the region after a Russian nuclear first use—in which case NATO's entire deterrent posture would be called into question in far more fundamental ways than merely Baltic relations—it seems unlikely that Moscow would welcome a lower nuclear threshold.

Another factor that could serve to undercut the strength of this variable over time is trends in U.S. political support for international commitments. The current U.S. administration has broadcast a somewhat mixed message about the desirability of the U.S. role in formal alliances, particularly if other members are viewed as not contributing enough. To the extent that these views are characteristic of a long-term trend in U.S. public and official attitudes, other countries will have greater reason to question the automaticity of U.S. responses to any challenges to major alliances.

Variable 3: The Degree of Actual and Perceived Credibility of Political Commitment to Fulfill Deterrent Threats

Measurements: Existence of alliance with local partner, deployment of local forces with a clear trip-wire function, unqualified public statements and commitments that engage national honor and prestige, legislative actions (resolutions), scope of intramilitary partnership with client state, amount and type of training, arms sales and transfers to client state as signal of commitment.

The decision taken at the 2016 NATO Summit in Warsaw to deploy four multinational battle groups signals the Alliance's commitment to engage forces and establish a durable military presence in Eastern Europe. These battle groups—deployed in Estonia, Latvia, Lithuania, and Poland on a rotational basis—will also function as a trip wire for eventually advancing Russian troops.[97] U.S. military assistance to beneficiaries of ERI has increased over the years—as did ERI as a whole—from $41 million in FY2015 to $218 million in FY2018 for additional bilateral and multilateral exercises

---

[97] North Atlantic Treaty Organization, "Boosting NATO's Presence in the East and Southeast," August 11, 2017d. For a critique of the trip-wire function of these forces, see Dianne Pfundstein Chamberlain, "NATO's Baltic Tripwire Forces Won't Stop Russia," *National Interest*, July 21, 2016.

and training, and from $14 million in FY2015 to $267 million in FY2018 for building partnership capacity, showing an escalating investment and commitment on the part of the United States.[98]

Some European allies might show less of a willingness to go to war for a NATO ally, at least at the public opinion level. A Pew Research Center survey conducted in early 2015 showed that, at the time, the perception in Europe that the United States would use military force to defend a NATO member attacked by Russia was high, with more than 65 percent of respondents in Canada, France, Germany, Italy, Spain, and the United Kingdom expressing that view. The only country represented in the survey that had more mixed views was Poland, where only 49 percent of respondents trusted the United States to intervene militarily (31 percent believed the United States would not use military force).[99] The same poll, however, showed limited support from several NATO members for collective defense, with a majority of respondents in France, Germany, and Italy not supporting the use of military force to assist a fellow NATO member against Russia.[100] Assuming that an actual Russian invasion of the Baltics will not fundamentally change these opinions, results from this poll suggest that any decision by the leadership of these countries to spend military resources defending the Baltics could have high political costs domestically, a factor that may constrain their response. Yet this lack of support for military action in support of an ally on the part of several European countries does not necessarily constrain the action of NATO, which does not have to be based on unanimous decisions. If Russia is convinced that the United States is committed to responding militarily to an attack on the Baltic States, the deterrent message will be credible. As of late 2017, the Russian leadership and elite seemed to believe that an aggression in the Baltic States would be met with a military response.[101]

As suggested above, Russian nuclear posturing could affect the credibility of a potential NATO response. Moscow might calculate that threatening nuclear escalation could fragment any political consensus behind NATO's response to aggression. There is every reason to expect Russia to leverage the coercive power of its nuclear arsenal, as well as many nonmilitary informational tools, to achieve such outcomes. Again, however, this is something of a dilemma for Russia: If it undertakes actual aggression,

---

[98] U.S. Department of Defense, Office of the Under Secretary of Defense (Comptroller), *European Reassurance Initiative: Department of Defense Budget, Fiscal Year (FY) 2016*, Washington, D.C.: U.S. Department of Defense, February 2015, pp. 14–15; U.S. Department of Defense, Office of the Under Secretary of Defense (Comptroller), *European Reassurance Initiative: Department of Defense Budget, Fiscal Year (FY) 2018*, Washington, D.C.: U.S. Department of Defense, May 2017, pp. 23, 25.

[99] The survey question referred explicitly to "neighboring countries" of Russia (and NATO allies) as the potential targets. See Katie Simmons, Bruce Stokes, and Jacob Poushter, "NATO Publics Blame Russia for Ukrainian Crisis, but Reluctant to Provide Military Aid," Pew Research Center, June 10, 2015, p. 54.

[100] Simmons, Stokes, and Poushter, 2015, p. 53.

[101] Frederick et al., 2017, pp. 59–60.

it will trigger widespread NATO military responses that then make it much more difficult for the Alliance to disengage and back down.

Long-term political trends could also affect the political viability of U.S. and allied deterrent threats. Both in the United States and in Europe, those trends are reflecting a greater influence of nationalist and populist sentiments and parties. To the extent that they generate a more skeptical attitude toward international commitments, the political will underpinning NATO Article V promises could ebb. This has not happened yet; indeed, the trends over the last several years have been in the opposite direction. But questions about foreign commitments, including deterrence of Russian aggression in Eastern Europe, are increasingly being raised on both sides of the Atlantic.

Variable 4: The Degree of National Interests Engaged in the State to Be Protected
Measurements: History of relationship, perceived geostrategic significance, measurable economic interests, statements of strong interests in the deterrer's public debate, evidence of perception of degree of interests by the potential aggressor.

Historically, the United States has enjoyed good diplomatic relations with the Baltic States. Estonia, Latvia, and Lithuania did not inherit any military forces from the Soviet Union and had to build their defenses from scratch, which they did with U.S. support. The United States also supported the Baltic States' efforts at regional defense integration that resulted in the creation of, among other structures, the Joint Baltic Peacekeeping Battalion, the Joint Baltic Naval Squadron, the Baltic Air Surveillance Network, and the Baltic Defense College.[102] The Baltic States were invited to join the Alliance during the NATO Summit in Prague in 2002, a decision in which the United States played a "decisive" role.[103] The Baltic States have been strong supporters of the United States not only diplomatically but also militarily, contributing to the war efforts in Afghanistan and Iraq.[104] Their geostrategic significance, for the United States, lies mostly in their membership to NATO and the fact that an unchecked aggression against them would undermine the Alliance. Since the commitment to NATO—which was reiterated by the current U.S. administration—supports the U.S. global strategic posture and makes the United States a credible ally, membership in the Alliance elevates the importance of the Baltic States for U.S. national interests. Or, as Vice President Pence put it during his July 2017 trip to Estonia, "America has no small allies."[105]

---

[102] F. Stephen Larrabee, *The Baltic State and NATO Membership*, Santa Monica, Calif.: RAND Corporation, CT-204, 2003.

[103] U.S. Embassy in Estonia, "U.S.-Estonia Relations," n.d.

[104] Heather Conley, "The Baltic States in the World," Remarks to the Baltic American Freedom League, Los Angeles, April 24, 2004.

[105] Biden, 2017.

Accordingly, U.S. leaders' frequent visits to the Baltics, both under the Obama and Trump administrations, suggest that the region is a U.S. priority or, at the very least, that U.S. officials want it to be seen that way.[106] The Russian media have duly noted and reported these high-profile visits to the Baltics.[107]

Conclusion

Strength of the U.S. Deterrence Posture in the Baltics

The application of the framework developed in Chapter Three to the case of the Baltic States makes apparent some strengths and weaknesses of the U.S. and NATO deterrent message to Russia. Our color-coding of each of the 12 variables from the framework in Table 4.1 is notional, and ultimately reflects an analytical judgment rather than the outcome of a model. We have color-coded a variable in green if it indicates strong deterrence, in orange if it indicates mixed deterrence, and in red if it indicates weak deterrence.

Russia's motivation to attack the Baltics appears low overall, suggesting that even a mixed deterrence posture on the part of the United States and NATO could be sufficient to prevent a Russian attack. Russia is dissatisfied with the current strategic environment, but there is no indication that the Baltic States are part of a plan to change that situation. Russia does not appear to fear a sudden change in the strategic situation that would warrant it to increasingly engage in risk-taking. It is not locked into a course of action that would leave it with no option besides aggression. Finally, its level of national interest in the Baltic States is fairly low, even with the presence of Russian minorities in Estonia and Latvia.

The second category of variables shows that the United States is mostly clear and explicit regarding what it seeks to prevent and what actions it would take in response. The United States and NATO have made clear that they will respond to a Russian military incursion in the Baltic States, even though they have been less clear about their intentions in the event of an unconventional operation on the part of Russia. The type of response that they will consider is not clearly spelled out, but the deployment of military forces makes it obvious that military solutions will be on the table—although

---

[106] Examples of U.S. decisionmakers who have visited the Baltics since 2014 include President Barack Obama; Vice President Joe Biden; Vice President Mike Pence; Senators Thad Cochran, Alexander Lamar, and John McCain; Speaker of the U.S. House of Representatives Paul Ryan; and Defense Secretary James Mattis. On the case of Estonia, see Andrew Hanna, "How a Tiny Baltic Nation Became a Top Destination for U.S. Officials," *Politico*, July 29, 2017.

[107] See, for example, "Peskov: RF bespokoit rasshirenie al'iansov k ee granitsam, a ne otnosheniia sosedei s SSHA" [Peskov: Russia is concerned about the enlargement of alliances to its borders, but not about relations of its neighbors with the U.S.], August 1, 2017; and "Spiker Kongressa: SSHA podderzhat Estoniyu v otnoshenii ugroz s vostoka" [Speaker of Congress: USA will support Estonia with regard to threats from the east], April 22, 2017.

Table 4.1
Application of the Deterrence Framework to Russia in the Baltic States

Category	Variable	Level
How motivated is Russia?	General level of dissatisfaction with status quo and determination to create a new strategic situation.	High in general, but low in relation to the Baltic States. Russia appears to have made its peace with the NATO membership of the Baltic States, which it does not consider as part of its "zone of influence," like Georgia or Ukraine.
	Degree of fear that the strategic situation is about to turn against Russia in decisive ways.	Low. While Russia resents U.S. military programs and their implications for its own defense, as well as NATO's enhanced defense posture in Eastern Europe, nothing indicates that it perceives the situation as about to radically worsen.
	Level of national interest involved in specific territory of concern.	Low. Russia has no economic interests in the Baltic States, who cannot be military allies since they are NATO members. Russia's main interest in the Baltic states relates to their Russian populations.
	Urgent sense of desperation or requirement to act; whether aggressor is locked into course of action.	Low. Putin's approval ratings suggest that he has ample political leeway to choose among various options. Russia has not committed to any specific course of action with regard to the Baltic States.
Is the United States clear and explicit regarding what it seeks to prevent and what actions it will take in response?	Precision in the type of aggression the United States seeks to prevent.	High. While cyber and unconventional attacks may or may not trigger Article V, the United States and NATO have made clear that a military aggression of the Baltic States would do so.
	Clarity in the actions that will be taken in the event of aggression.	High. While the United States and NATO appear reluctant to commit to military action in principle, their military deployments in Eastern Europe clearly indicate to Russia that they are seriously considering military options.
	Forceful communication of these messages to outside audiences, especially potential aggressor(s).	High. Frequent messaging by high-level officials, clearly addressed at Russia.
	Timely response to warning with clarification of interests, threats.	High. Invasion of Crimea was followed by clear military support to the Baltic States. Russia's subsequent moves in the Baltics have remained at such a low level that they probably do not warrant a clarification, or even a simple reiteration, of the threat.
Does Russia view U.S. threats as credible and intimidating?	Actual and perceived strength of the local military capability to deny the presumed objectives of the aggression.	Low. Even with NATO's reinforcements, local forces remain much inferior to Russia's.
	Degree of automaticity of U.S. response, including escalation to larger conflict.	Medium. A military response is not automatic, and the United States and its allies can decide on the most appropriate response if Russia were to attack. Yet the presence of ground forces will act as a trip wire in the event of a Russian attack and make conflict inevitable.

Table 4.1—Continued

Category	Variable	Level
	Degree of actual and perceived credibility of political commitment to fulfill deterrent threats.	High. U.S. political commitment has been reaffirmed, and Russia appears to believe that the United States would respond militarily to an attack against the Baltic States—with or without the support of its least supportive European allies.
	Degree of national interests engaged in state to be protected.	High. The main value of the Baltic States for the United States lies in their being part of NATO. The commitment to NATO supports the U.S. global strategic posture and makes the United States a credible and desirable ally.

they might not be automatically chosen. The U.S. commitment to collective defense has been reiterated many times by high-level officials who are clearly communicating to Russia as much as to their local audiences. Finally, the U.S. commitment to collective defense was made clear after the invasion of Ukraine. While it does not appear to have been reiterated whenever Russia has attempted provocations in the Baltic States, this might have simply been a way for the United States to downplay Russian actions and prevent minor incidents from escalating.

Finally, the question remains as to whether the defender's threat is credible in this case. This third category scores the lowest in terms of strength of the U.S. deterrence posture. The actual and perceived strength of the local military capability to deny the presumed objectives of the aggression is low, as NATO forces deployed currently in Eastern Europe are no match for the Russian military. While a military response is not automatic, the presence of ground forces acting as trip wire effectively makes a conflict inevitable were Russian forces to attack. U.S. political commitment has been reaffirmed, suggesting the United States would oppose militarily a Russian incursion in the Baltic States even if some NATO members were to prefer a nonmilitary solution. Finally, the degree of national interests engaged in the Baltic States is high, since abandoning them would cause irreparable damage to the credibility of the Alliance.

One important implication of these variables is to reduce the potential for a fait accompli—a Russian attack that seizes all or most of the Baltics, and this attack is compartmentalized from the larger European military situation. While the current local military balance is not favorable to NATO, recent improvements in NATO's posture throughout the region, better planning and coordination among NATO members, and repeated statements of the U.S. and NATO commitment to the defense of the Baltics all suggest that any Russian attack in the Baltics would trigger a wider war whose consequences and outcome would pose the most profound risks of national survival for all sides. One of the messages of our framework is that an unfavorable local balance of forces *alone* need not undermine the strength of deterrence, if many other variables continue to reinforce the potential risks and costs of aggression.

Even with mixed credibility of the local force balance, therefore, the U.S. and NATO deterrent posture may be sufficient to hold back Russia given Russia's limited ambitions in the Baltics. As noted above, Russia has significantly more forces to call upon in the local area. They would likely seek a rapid resolution of any Baltic conflict, trying to achieve a quick fait accompli before NATO could adequately react. The action would pose NATO with the dilemma of either accommodating Russia's possession of the Baltics or undertaking a long and extremely costly campaign to eject Russian forces and win them back. From the standpoint of some measures of classic deterrence by denial, therefore, NATO's posture continues to betray some weaknesses.

Yet our research into the character of deterrence suggests that these must be balanced against other variables that help determine the success or failure of deterrence, many of which demonstrate much greater health than in cases (such as Kuwait in 1990) where deterrence of interstate aggression failed. One is the broad area of aggressor motivation: Russia may be motivated to attack the Baltics simply to test NATO's commitment to defending its members rather than because it sees any intrinsic value in the Baltic States, but there is no evidence at this point that it would be willing to take such a risk to achieve this outcome.[108] The Baltic States lack the characteristics that made the annexation of Crimea relatively painless overall,[109] and if NATO were to respond militarily, Russia would find itself choosing, as Frederick and colleagues put it, "between devastating conventional defeat or an even more devastating escalation to nuclear confrontation."[110] While Heather Conley and colleagues assert that "President Vladimir Putin has shown an increasing readiness to take significant risks,"[111] the Russian leadership has also shown to be sensitive to costs when undertaking military action.[112]

One would have to assume a Russia extremely willing to engage in risk-taking for such a scenario to be plausible—or the occurrence of events that would make its leadership much more prone toward risk-taking than it has shown to be so far. The nature of the risks involved, in fact, are a key aspect of NATO's deterrent advantages in the Baltics, which differentiate the situation from that of a number of other recent cases—such as gray-zone aggression against Ukraine leading to partial conventional aggression—in which deterrence has failed. The Baltic States are NATO members,

---

[108] Additionally, assuming this was Russia's intent, it would not have to target the Baltic States to test NATO, but could pick another NATO ally.

[109] Kofman et al., 2017, note that "Russian leaders are likely to consider Crimea an operation that could not be easily repeated elsewhere and Eastern Ukraine to be a strategic success but an unsuccessful operation" (pp. xiii–xiv), and further conclude that there is nothing "to suggest that the Russian military sees the utility of a Crimea- or Eastern Ukraine–type approach against a NATO Member" (p. 77).

[110] Frederick et al., 2017, p. 2.

[111] Conley et al., 2016, p. 2.

[112] Kofman et al., 2017, pp. 62–63.

and border other NATO states, and U.S. presidents have repeatedly indicated that the United States would lead a NATO response to any aggression against them. NATO has undertaken significant initiatives since 2014 to enhance its capacity for war fighting on the Continent. All of these aspects differentiate the situation in the Baltics from other deterrent cases.

It is important to stress that this analysis assumes the current political context in Europe and the United States—or something close to it. While it has come to reflect greater degrees of nationalist and populist sentiments, and while tensions have emerged especially between Europe and the United States on a number of issues, the basic commitment to NATO security guarantees, reflected in part by rising investments to bolster deterrence, remains strong. If this were to change—if the political trends behind a more skeptical attitude toward foreign commitments were to accelerate—many of the findings in this report would have to be reevaluated. A number of the variables we have coded as healthy would become increasingly ambiguous, and the overall status of deterrence would weaken.

Ways to Enhance Deterrence Posture in the Baltics

For each variable that is not coded as green in Table 4.1, there are steps that the United States and/or its army could take to strengthen the U.S. deterrent message. Measures to improve the credibility of the U.S. deterrent message could include continued investment in military capabilities to be forward deployed in the Baltics in an effort to further tip the local military balance in the direction of the Alliance. Several studies have already charted possible courses of actions to achieve that objective.[113] The United States might also want to make its response more automatic, in order to leave Russia with no doubt that an attack will elicit a military response in all cases. While the tripwire function of the ground troops currently deployed plays this role to some extent, additional measures could include further reinforcing and reiterating the U.S. commitment to NATO to make it clear that the United States will be part of a military response if Russia decides to attack the Baltic States. Table 4.2 summarizes these different options to enhance the U.S. deterrent posture in the Baltics.

To be sure, the United States would also benefit from addressing some of the variables currently coded in green in Table 4.1. For instance, it could entice Estonia and Latvia to better address the needs of Russian-language speakers. Yet this would likely not increase or decrease Russia's motivation to attack the Baltics. Rather, it would diminish Russia's ability to influence the Baltic States, which would be in itself a success for the United States but have little to no impact on its deterrence posture.

As suggested in previous chapters, the United States could conceive of its strategy for enhancing deterrence in one of two ways. It could start a process of steps designed to eventually provide local military superiority, or at least equivalence, within

---

[113] See Conley et al., 2016; Kramer and Craddock, 2016; and Reed et al., 2016.

Table 4.2
Recommendations for U.S. Deterrent Posture in the Baltics

Category	Variable	Policy Options for Enhancing Deterrence
How motivated is Russia?	General level of dissatisfaction with status quo and determination to create a new strategic situation.	1. Engage with Russia to the degree possible to address concerns about strategic priorities and vulnerabilities including issues specific to Baltics; coordinate NATO-EU process to do same. Until the situation in Ukraine is resolved, this will have to start with small initiatives.
	Degree of fear that the strategic situation is turning against it.	2. Pursue specific confidence-building measures around military deployments, exercises, etc., including discussions on regional arms control accords. 3. To the degree possible, gradually restart programs and processes to develop personal relations between senior officials and military leaders where possible and promote regular dialogue. 4. Avoid deployment of most escalatory capabilities in region, including strategic strike systems.
	Level of national interest involved in specific territory of concern.	*Options above, plus:* 1. Take political/economic measures to reduce Russian leverage in Baltics: Engage Estonia and Latvia to address fears of local Russian-language populations; promote and support diversification of energy supplies in the Baltic States.
	Urgent sense of desperation, need to act.	*Options above, plus:* 1. Develop strong sources of intelligence to gauge Russia's perceptions and anticipate escalatory behavior.
Is the United States clear and explicit regarding what it seeks to prevent and actions it will take in response?	Precision and clarity in the type of aggression the United States seeks to prevent.	1. Continually reaffirm U.S. commitment to Article V status of the Baltics. 2. Be explicit in public statements about commitment to take some actions in response to conventional aggression (even if sustaining some ambiguity on the form of response to low-level gray-zone activities).
	Clarity and consistency in the actions that will be taken in the event of aggression.	1. Reiterate public statements of commitment to defend the Baltics with full weight of NATO capabilities. 2. Have military leaders reiterate public statements of war planning for such a scenario, describing theater-wide capabilities that will be brought to bear. 3. Conduct exercises and training to prepare NATO for the scenario.
	Forceful communication of these messages to outside audiences, especially potential aggressor(s).	1. Have U.S. and NATO officials reiterate these messages in direct communication with Russian leaders including President Putin. 2. Include reaffirmations of commitment and military planning basis in NATO communiqués, national security strategies, and other strategy statements. 3. Include theme in speeches and articles by senior U.S. and NATO officials.
	Timely response to warnings with reiteration of threats.	1. In case of Russian massing of forces in exercises or other ways that would provide potential for short-notice aggression, reiterate commitments to Baltic defense.

Table 4.2—Continued

Category	Variable	Policy Options for Enhancing Deterrence
Does Russia view the U.S. threats as credible and intimidating?	Actual and perceived strength of the defender's local military capabilities and its ability to defeat possible aggression.	1. Take steps to address gaps in U.S. Army capabilities for combined arms operations in Europe. 2. Address shortfalls or gaps in U.S. Air Force and U.S. Navy capabilities for European campaign. 3. Deploy modest additional heavy ground forces to region—Baltics, Germany, and Poland. 4. Continue to work with Baltic States to enhance local defensive capabilities, including whole government efforts to enhance societal resilience and asymmetric resistance techniques. 5. Address issues of precision weapons supply for potential conflict. 6. Continue to enhance command and control and logistical foundations for a campaign.
	Degree of automaticity of U.S. response, including escalation to larger conflict.	1. Publicize elements of war plan that integrate far-flung air, maritime, cyber, and other capabilities in early hours and days to convey inevitable and immediate spread of conflict. 2. Publicize autonomy given to local commanders in requesting and/or deploying capabilities from outside theater (short of nuclear assets).
	Degree of actual and perceived credibility of political commitments to fulfill threats.	1. Make public case in U.S. for commitment; drive polling numbers on issue. 2. Demonstrate solidarity of legislative branch on commitment: congressional delegations, resolutions.
	Degree of U.S. national interests engaged in state to be protected.	1. Take steps to boost U.S. economic investments in the Baltics. 2. Conduct high-profile U.S. leadership visits to Baltics, including visits from the president.

the Baltic region itself. This approach would ultimately demand very large deployments of heavy combat capability in the region, including U.S. and NATO armored brigades, substantially expanded local logistical and operational command capabilities, enhanced long-range fires and short-range air defense for the local fight, and nearby tactical air deployments. Early investments under the NATO European Deterrence Initiative could lay the groundwork for these goals by building facilities in the Baltics to host and sustain such a force, for example.

On the other hand, the United States and NATO could decide, at least as a short-term step, to focus on introducing new complications to Russian planning for a potential Baltic conflict without necessarily seeking to create comprehensive local denial capabilities. This approach would reflect the blended notion of denial and punishment outlined in Chapter Three. It might focus on such investments as limited but symbolically important NATO ground force deployments in the Baltics; expanded and regular exercises; investments in the Baltic armies' capabilities for long-term, population-based resistance; more limited logistical capabilities to host rotational units as signaling devices; enhanced regional logistics and command facilities in places such as Germany and Poland; improved long-range fires; improved maritime strike capabilities; and expanded and more resilient air bases capable of hosting strikes into the Baltic region.

Obviously, these options are not mutually exclusive. Everything in the second approach could complement the ability of larger heavy forces to defend the Baltics. But the two approaches might suggest significantly distinct early investment priorities.

Overall, the current deterrent posture of the United States in the Baltic States is one that is reasonably strong, as it combines a clear and relatively credible message with low motivation to attack on the part of the would-be aggressor. This posture can still be improved through a stronger defense posture in the Baltic region, as well as a reiteration and reinforcement of U.S. commitment to NATO and collective defense. This stronger defense posture, however, should be designed to make clear that it only aims at defending the Baltic States. If it were to be perceived by Russia as an existential threat, this would weaken the U.S. and NATO deterrent posture by increasing Russia's motivation to attack, and raise the risks of military conflict.

CHAPTER FIVE

Conclusions, Recommendations, and Implications for the U.S. Army

This study has examined the sources of success for extended deterrence of interstate aggression. It has done so through a survey of existing research, including quantitative analyses; conducting a new quantitative assessment of extended deterrence cases since 1945 as measured against a number of potential explanatory variables; and the assessment of four detailed case studies—West Berlin in the early Cold War, the Nordic countries during the Cold War, the Iraqi invasion of Kuwait in 1990, and the conflict in the Republic of Georgia in 2008. The study has identified a set of 12 variables, listed in Table 5.1, that our research has confirmed as being associated with the success or failure of extended deterrence.

Beyond validating that framework, our research highlights specific themes related to successful extended deterrence. They include the following:

- Potential aggressors' motivations are highly complex and typically respond to many variables whose interaction is difficult to anticipate.
- Opportunism in aggression seems generally less common than desperation through paranoia about growing threats to security or status. Large-scale aggression tends to emerge as a last resort based on intense fears rather than an opportunistic grab, though there are certainly cases of the latter.
- Clarity and consistency of deterrent messaging is essential. Half-hearted commitments to allies risk being misperceived.
- The importance of the principle of "firm but flexible": leaving an adversary without a way out is not an effective way to sustain deterrence. Compromise and concession are typically part of any version of successful extended deterrence of large-scale aggression to help meet a potential aggressor's interests and deprive it of a sense of imminent threat that would mandate conflict.
- Multilateral deterrent contexts are especially dangerous. Deterring an aggressive ally *and* an opposing major power at the same time is extremely difficult.

In sum, this analysis supports a specific view of the requirements of extended deterrence of interstate aggression. It suggests that *the foundation for deterrence outcomes—*

Table 5.1
Conclusions: Key Variables Governing the Success of Deterrence

Category	Variable
How intensely motivated is the aggressor?	1. General level of dissatisfaction with status quo and determination to create a new strategic situation.
	2. Degree of fear that the strategic situation is about to turn against the aggressor in decisive ways.
	3. Level of national interest involved in specific territory of concern.
	4. Urgent sense of desperation, need to act.
Is the defender clear and explicit regarding what it seeks to prevent and what actions it will take in response?	1. Precision and consistency in the type of aggression the defender seeks to prevent.
	2. Clarity and consistency in the actions that will be taken in the event of aggression.
	3. Forceful communication of these messages to outside audiences, especially potential aggressor(s).
	4. Timely response to warning with clarification of interests, threats.
Does the aggressor view the defender's threats as credible and intimidating?	1. Actual and perceived strength of the local military capability to deny the presumed objectives of the aggression.
	2. Degree of automaticity of defender response, including escalation to larger conflict.
	3. Degree of actual and perceived credibility of political commitment to fulfill deterrent threats.
	4. Degree of national interests engaged in state to be protected.

that is, the initial and in some ways decisive variable—is provided by *aggressor motivations*. The most obvious but also most important fact about extended deterrence is simply that weakly motivated aggressors are easy to deter; intensely motivated ones, whose level of threat perception may verge on paranoia, can be impossible to deter. The first step toward bolstering deterrence is therefore to manage the motives of a potential aggressor; doing so often requires concessions, as well as steps to shape the surrounding geopolitical context to ease its concerns and also raise the political costs of aggression.

This finding supports the broader definition of deterrence suggested in Chapter Two. What the United States is really trying to do in these cases, in the wider sense, is *dissuade* a potential aggressor from violent action. Dissuasion includes threats of what the United States (and others) will do in response. But it also includes policies designed to ease fears that might seem to require aggression and to shape the overarching geopolitical context to make aggression both unnecessary and counterproductive.

This analysis then suggests that *clarity in what is to be deterred, and how the United States will respond if deterrence fails,* is the second essential element of a successful deterrent posture. Lack of clarity invites opportunistic aggression and provides fuel for wishful thinking for highly motivated aggressors. There are no identifiable cases of failed extended deterrence, on the other hand, in which the United States was entirely clear in its interests and intentions. In most cases, the United States has fortified that clarity with concrete policies such as alliances, repeated senior-level reiterations of U.S. promises, military exercises and training programs, and deployment of at least symbolic military forces. Clarity backed up by concrete evidence of commitment is a cornerstone of effective deterrence.

Too often, effective deterrence is equated with the deployment of military forces or capabilities that achieve some specified correlation of forces in a contested area. Such balances are surely important, and the absence of a meaningful war fighting capability has invited aggression in certain cases. But even in those cases, the capability gap appears to have been less important in itself than for what it implied: the absence of a clear U.S. commitment to respond to aggression. North Korea invaded South Korea in 1950 and Iraq attacked Kuwait in 1990 not merely because of an absence of U.S. forces able to win the local fight. They attacked because they believed the United States would not respond—and saw the absence of a large-scale commitment as a notable but secondary source of evidence for that weak commitment.

Effective deterrence, then, is not *merely* about the creation of sufficient local capabilities to win a conflict at the scene of an attack. It is about a combination of factors working together to shape the motivations of a potential aggressor: shaping the geopolitical context, furnishing clarity in U.S. promises, and backing up that clarity with sufficient local forces to deny an aggressor the possibility of an easy victory and to provide strong confirmation of the U.S. commitment at stake. Deterrence—or the broader task of dissuasion—is thus first and foremost a diplomatic and geopolitical challenge, and secondarily a military one.

We then applied these findings, and in particular the key deterrence variables, to the current situation in the Baltics to assess the health of the U.S. and NATO policy of deterring Russian aggression against those states. Broadly speaking, in terms of Russia's degree of motivation to undertake aggression and the level of U.S. clarity and commitment, as well as many other of the variables from the set of the leading 12, we find that the current U.S. and NATO policy meets many of the criteria for successful deterrence. The criteria do suggest areas where it could be improved, however.

Recommendations

In its focused application, this analysis has dealt primarily with one case study—the Baltic region. It also carries general implications for the practice of deterrence more broadly. We have therefore organized the recommendations below in two tiers—a handful of overarching recommendations for the conduct of extended deterrence, followed by specific recommendations for implementing those broad principles in the Baltic case.

Principle One: The United States should carefully assess the national interests and motives of any potential aggressor and seek to ease security concerns that could lead to aggression. This recommendation is essentially to implement the "firm but flexible" approach to deterrence described above, working to reduce urgent security fears that could lead to aggression through desperation.

In the case of dealing with Russia in the Baltics, this recommendation points to a number of specific policies and actions:

- In order to set the context for more specific policies, the United States should bolster efforts to engage Russia at senior levels and discuss steps each could take to ease tensions and create mechanisms to avoid security concerns.
- The United States should avoid deployment of the most provocative systems in or near Eastern Europe, including strategic strike assets.
- The United States should begin discussions with Russia on a possible agreement to thin forces on both sides of the Russian-Baltic borders, or even on a broader successor to the Conventional Forces in Europe Treaty, to enhance the security of the Baltics while reducing the deployment of U.S. and NATO troops on Russia's borders.

Principle Two: The United States should work diplomatically to create a geopolitical context hostile to aggression. A potential aggressor's motives are crucially shaped by the surrounding context and a sense of likely political and economic consequences of aggression. In all such cases, the United States can use regional diplomacy to generate clear commitments of response to aggression.

In the Baltic case, this recommendation suggests a number of specific approaches:

- Continual efforts to strengthen the NATO alliance's political foundations and sustain general relations among its members. This includes unqualified statements of continued U.S. support for the Alliance.
- Work with fellow NATO members to reiterate public commitments to Article V countries and a determination to respond to aggression.
- Working beyond NATO, with countries such as Brazil, India, and Japan, to reinforce the norm of nonaggression and the mutual commitment to respond powerfully to cases of large-scale military aggression.

Principle Three: The United States should seek clarity in the actions it is pledged to deter and the general scope of its promised reaction to aggression. From the standpoint of U.S. policy, extended deterrence rests most significantly on a foundation of clarity of intent. The United States should be specific and clear about actions it is committed to deter and likely actions it will take if that commitment is challenged.

In the Baltic case, this recommendation suggests a number of specific approaches. As noted in Chapter Four, the United States and NATO gain some degree of flexibility from a bit of ambiguity in precisely how they would respond to various forms of aggression. But it should be absolutely clear *that* they would respond, and there should be an effort to improve Russian awareness of various means the United States and NATO have at their disposal in terms of *how* to respond.

- The United States should reaffirm the Article V status of the Baltic States and its commitment to respond to aggression against NATO members.
- Without necessarily specifying a rigid response mechanism, the United States should outline options that would be available to be taken in the event of aggression against NATO members, including military response to restore the status quo ante, as well as large-scale political and economic consequences.

Principle Four: The United States should take specific steps that reinforce key criteria for successful deterrence as outlined in this study's framework. This study highlights 12 variables as being closely associated with the success or failure of extended deterrence. In each case, the study identifies specific action that can reinforce the deterrence-supporting elements of each of those variables. This analysis provides a detailed menu of options to bolster deterrence in specific cases.

In terms of the Baltic example, Chapter Four offered a number of specific examples of such steps across the 12 major criteria in the framework of variables governing deterrence success. Many of these are embedded in the other principles, but additional steps include:

- Visits and statements of support from senior U.S. leaders, including the president.
- Strengthening and diversifying relationships with the Baltic countries, including promoting new economic investment from the United States and NATO members, as well as continued diplomatic engagements.
- Measures to strengthen communication and crisis management avenues both within NATO and between NATO and Russia in order to prevent accidents and manage any escalatory situations that do emerge.

Principle Five: The United States should deploy or support sufficient local capabilities to signal the seriousness of its commitment, to deprive an aggressor of a possible fait accompli, and to offer enough defensive power to assure that the conflict will not remain limited—without employing specific capabilities or postures that the other side will view as immediately threatening to its security. As we have argued above, a demonstrated ability to win the local fight is not necessary for deterrence—but a strong enough capability to forestall a fait accompli is still a useful complement to deterrence. The local force objective should be to make it "messy enough" to deprive any thought of an easy, quick, isolated win. The challenge is how to measure this threshold—and to ensure that the new capabilities do not create the provocation dynamics that the "firm but flexible" approach is designed to avoid.

In the Baltic case, this recommendation suggests a number of specific approaches:

- The United States should work with NATO allies to fund new defensive capabilities for the Baltic States, potentially including large numbers of portable antitank weapons, mines capable of being rapidly laid in the event of crisis or war, short-range rocket systems, swarming drone technology, and more.
- The United States should reinforce the existing NATO multinational battalions in and around the Baltics with additional U.S. heavy forces.
- The United States should invest in needed capabilities for successful combined arms operations in Europe—especially in ground force capabilities that have been identified by the U.S. Army as key gaps.

Implications for the U.S. Army

This portrait of the factors governing success and failure in extended deterrence of interstate aggression carries a number of implications for the U.S. Army. These include the following:

- Ground forces have a critical role to play in sustaining deterrence in general and in Eastern Europe in particular. They serve as a powerful signal of U.S. commitment and resolve and underline the clarity of U.S. threats.
- However, effective deterrence—dissuasion of aggression—demands a comprehensive integration of instruments of national power, beginning with diplomacy and negotiations. Ground forces should be conceived of as a supporting element to a larger strategy of deterrence, not the basis of the policy themselves.
- The ability of local U.S. forces to win a contest outright is of less importance than the presence of some forces, wider steps to bolster deterrence, and a minimum ability to forestall defeat to assure that the conflict will remain contained.
- Nonetheless, limited additional U.S. heavy forces in the region, with some deployed in the Baltic nations themselves, would help rule out the potential for any rapid and limited strike by Russian forces.
- The United States should also make the necessary investments to close key gaps in U.S. Army capabilities for combined arms operations in Europe, including long-range fires, new systems for operating in an electronic warfare environment, short-range air defense, antiarmor weapons, and more.
- Special Operations Forces (SOF) and train-and-advise forces can play an important role in enhancing the defensive capabilities of partner nation forces. Investments in defensive capabilities in partner countries can be more cost-effective than deploying U.S. forces.
- Ground forces provide multiple options to enhance deterrence while sidestepping the provocative, deterrence-threatening risks of some other systems, such as strategic strike capabilities.

APPENDIX A

Quantitative Analysis: Cases of U.S. Extended Deterrence Since 1945

As noted in Chapter One, we surveyed the literature on deterrence and derived the framework of 16 factors that help determine extended deterrence success. Many of these are grounded in extensive empirical work, including some detailed quantitative studies, but we nevertheless sought to confirm the utility of these factors with additional research and analysis.

The first such step was to perform a quantitative assessment of extended deterrence cases since 1945. In order to explore and apply the range of variables developed in the framework in Chapter Two, we analyzed and coded 39 cases of U.S.-led extended deterrence since 1945. To select the deterrence cases where the United States employed military and/or economic pressure (in addition to political pressure) *to deter territorial aggression*, we reviewed a range of data sets that provide information on: U.S. deterrent interventions, the threats and imposition of sanctions,[1] militarized interstate disputes (MIDs) involving the United States since 1945,[2] and post-1945 interstate territorial claims from the Issue Correlates of War Project.[3] We used the U.S. deterrent interventions data set—the product of ongoing RAND research—as a base and searched for additional cases in the data on MIDs, territorial claims, and sanctions. For the sanctions data set, we looked into cases where "Issue" was coded as "Contain Military Behavior" or "Solve Territorial Dispute" in order to narrow the range of cases to assess.

Before coding cases, we organized them by determining whether they were instances of "immediate" or "general" deterrence. General cases refer to U.S. efforts to deter long-term or broader threats, whereas immediate cases are U.S. efforts in

---

[1] Clifton Morgan, Navin Bapat, and Yoshi Kobayashi, "The Threat and Imposition of Sanctions: Updating the TIES Dataset," *Conflict Management and Peace Science*, Vol. 31, No. 5, 2014, 541–558.

[2] Faten Ghosn and Scott Bennett, *Codebook for the Dyadic Militarized Interstate Incident Data*, Version 3.10, 2003.

[3] Bryan A. Frederick, Paul R. Hensel, and Christopher Macaulay, "The Issue Correlates of War Territorial Claims Data, 1816–2011," *Journal of Peace Research*, Vol. 54, No. 1, 2017, 99–108.

response to an immediate threat, and are often but not always a subcase of a longer, general deterrence case. For example, the U.S. Cold War deterrent posture in Europe case has four immediate deterrence subcases: the 1948 and 1961 Berlin crises, as well as the deployments of Jupiter and Pershing II missiles.

The diversity of cases—including geographic and temporal aspects, and which actors were involved—provided a broader lens through which to assess potential drivers of deterrence successes and failures. Of the 16 variables that the framework identifies, we coded five for the 39 selected cases:

1. How motivated was the aggressor?
2. How clear was the United States (both regarding what the United States wanted and the consequences)?
3. What was the local balance of forces?
4. What was the degree of U.S. interests involved?
5. Did the adversary believe the United States would respond?

We selected these five variables because they allowed us to roughly proxy each of the main categories in the framework, while still being plausible to code. Our interpretation of the first and fifth variables, in particular, intends to capture some factors from the framework that would be difficult to code consistently across a number of cases; for example, the literature addresses the potential impact of motivated reasoning and wishful thinking, but these factors would be difficult to distinguish in the cases from regular or wholesale "judgment." Therefore, they are folded into a broader assessment of "aggressor motivation" and "belief the United States will respond." The first variable—how motivated the aggressor was—sought to capture the range of factors explored in the framework, including the aggressor's level of satisfaction with the status quo, degree of fear that the strategic situation was about to change, level of national interest in the specific territory of concern, level of urgency or sense of desperation, degree of risk-accepting opportunism, and degree of wishful thinking. The final variable—whether the aggressor believed the United States would respond—serves as an aggregate and weighted assessment of the first four, and also captures framework variables such as the actual and perceived strength of the local military capability to deny objectives, degree of automaticity of U.S. response, degree of actual and perceived credibility of political commitment to fulfill deterrent threats, and reputation for resolve with potential aggressor.

We then coded whether deterrence succeeded or failed, in order to look for patterns that could shed light on the question of "what deters and why?" Four ongoing cases were not coded for success or failure, and so are excluded from the final tally of successful versus failed deterrence. In total, U.S. extended deterrence has succeeded more often than it has failed; we determined there were 24 cases where U.S. deterrence efforts succeeded and 11 cases where they did not.

Table A.1
All Cases

	Did deterrence succeed?	
Yes	24	
No	11	
Case still ongoing	4	

	How motivated was the aggressor?	Of these, how many cases of failed deterrence were there?
High	20	9
Medium	15	2
Low	4	0

	How clear was the United States?	Of these, how many cases of failed deterrence were there?
Clear	21	0
Somewhat clear	12	6
Ambiguous	6	5

	What was the local balance of forces?	Of these, how many cases of failed deterrence were there?
Clear U.S. and allied advantage	7	0
Ambiguous	24	8
Clear adversary advantage	8	3

	What was the degree of U.S. interests involved?	Of these, how many cases of failed deterrence were there?
High	25	3
Medium	12	6
Low	2	2

	Did the adversary believe the United States would respond?	Of these, how many cases of failed deterrence were there?
Yes	21	4
Uncertain	17	6
No	1	1

Table A.2
Cases of Failed Deterrence

	How motivated was the aggressor?
High	9
Medium	2
Low	0

	How clear was the United States?
Clear	0
Somewhat clear	6
Ambiguous	5

	What was the local balance of forces?
Clear U.S. and allied advantage	0
Ambiguous	8
Clear adversary advantage	3

	What was the degree of U.S. interests involved?
High	3
Medium	6
Low	2

	Did the aggressor believe the United States would respond?
Yes	4
Uncertain	6
No	1

The specific coding decisions connecting each case study with the variables were based on qualitative case research into each of the examples of extended deterrence. In many cases, the coding decisions were fairly binary and straightforward, such as cases where the United States had no clear deterrence statements at all (as in Iraq in 1990 or North Korea in 1950). In other cases, the coding demanded a more nuanced judgment where the evidence did not allow a clear and obvious judgment on the variable under consideration. In most cases we used a three-part scale—high, medium, and low—that allowed us to code the cases on a spectrum. Broadly speaking, cases that ended up having vague or conflicting evidence often ended up in the medium category,

Table A.3
Cases of Successful Deterrence

	How motivated was the aggressor?
High	10
Medium	11
Low	3

	How clear was the United States?
Clear	19
Somewhat clear	4
Ambiguous	1

	What was the local balance of forces?
Clear U.S. and allied advantage	7
Ambiguous	13
Clear adversary advantage	4

	What was the degree of U.S. interests involved?
High	19
Medium	5
Low	0

	Did the aggressor believe the United States would respond?
Yes	15
Uncertain	9
No	0

simply because there was no evidence to describe them as either high or low. However, as the results of the coding make clear, we were not forced to place the vast majority of cases into this middle ground; in many if not most instances, evidence was available to render a clearer judgment about whether a case did or did not reflect the variable.

This appendix offers some descriptive statistics of the findings of the case study analysis, and then delves into a few variables and outlier results in greater detail. Overall, the case studies generally support the findings of the literature review. U.S. clarity about desires and consequences, a clear U.S. and allied advantage in the local balance of forces, and low aggressor motivation were all associated with deterrence success.

Adversary belief or uncertainty as to whether the United States would respond was also strongly associated with U.S. extended deterrence success. There were, however, some outlier cases, including U.S. deterrence successes in the face of an adversary advantage in the local balance of forces and U.S. deterrence failures when U.S. interests were high and the aggressor believed the United States would respond.

Descriptive Statistics

Findings

Generally, the case study analysis conforms to the existing literature on deterrence. There were no deterrence failures when the United States made clear statements about both what it wanted and the consequences for noncompliance, when there was a clear U.S. and allied advantage in the local balance of forces, or when aggressor motivation was coded as low. Likewise, an adversary's belief that the United States would respond (15 of 24 cases) or uncertainty as to whether the United States would respond (nine of 24 cases), were also strongly associated with U.S. extended deterrence success. There were no deterrence successes when the adversary did not believe the United States would respond, nor when U.S. level of interest was coded as low.

However, while the overall patterns are in line with expectations, there were a handful of outlier cases (for example, deterrence successes in the face of an adversary advantage in the local balance of forces). The case studies also found that sanctions—absent military pressure—were generally not sufficient to deter territorial aggression.

Key Findings

In general, the case study findings align with those of the literature review, as outlined in Chapter Two. For example, our analysis showed no deterrence failures when

- the aggressor's motivation is coded as low
- the United States is clear—both about what it wants, and the consequences for crossing the United States
- there is a clear U.S. and allied advantage in the local balance of forces.

Additionally, deterrence success (24 total cases) was highly associated with cases where

- the United States is clear (19 of 24 cases)
- U.S. interests are high (19 of 24 cases; with no success cases when U.S. interests coded as low)
- the adversary believes the United States will respond (15 of 24 cases) or is uncertain as to whether the United States will respond (nine of 24 cases).

Deterrence failures (11 total cases) were associated with

- a highly motivated aggressor (nine of 11 cases); however, there were also ten cases of deterrence success in the face of a highly motivated aggressor (see Outlier Cases), suggesting high levels of aggressor motivation are not a sufficient factor on their own
- an ambiguous (five of 11 cases) or somewhat clear (six of 11 cases) United States
- an ambiguous local balance of forces (eight of 11 cases).

Outlier Cases

The overall patterns and findings conform to the general literature on extended deterrence; however, there are some outliers of interest. The United States had four extended deterrence successes in the face of an adversary advantage in the local balance of forces. The case studies also found that sanctions—absent military pressure—were generally not sufficient to deter territorial aggression. Three of the four cases with high U.S. interests and where the adversary did believe the U.S. would respond, but deterrence still failed, involved economic and political pressure but not military pressure. U.S. efforts to deter North Vietnam from intervening directly in South Vietnam from 1961 to 1964 made up the only case of deterrence failure in which U.S. interests were high, the United States applied military pressure, and the adversary believed the United States would respond.

In three of the 11 cases of deterrence failure, the degree of U.S. interest was coded as high. The United States had strong interests involved in trying to deter North Vietnam from intervening in South Vietnam from 1961–1964, Iraq from invading Kuwait in early 1990, and Soviet and North Vietnamese aggression against Laos, South Vietnam, and Thailand from 1955 to 1977 (via Southeast Asia Treaty Organization security guarantees). In the Iraq-Kuwait case, however, the United States faced a highly motivated adversary with an advantage in the local balance of forces, and did not clearly communicate deterrent threats and the adversary was uncertain as to whether the United States would respond.

In four of the 11 cases of deterrence failure, the adversary believed the United States would respond. However, three of the four cases involved only a combination of political pressure and sanctions (of threats thereof), revealing the limited utility of economic pressure as a tool to deter territorial aggression. Only the Vietnam War case involved the United States applying military pressure. This case may be worth closer investigation. In spite of strong U.S. interests and North Vietnamese belief that the United States would respond if North Vietnam intervened in South Vietnam, deterrence failed. The question remains, was North Vietnam simply willing to pay a high price, or was extended deterrence weakened by the fact that the United States was not completely clear on what the price would be?

There were also some surprising deterrence successes; for example, in ten of 24 cases of deterrence success, the aggressor was highly motivated (see Table A.4). These cases form a unique geographic cluster: eight of the ten cases occurred in East Asia (see Figure A.1). Four relate to Taiwan: a general deterrence effort to defend Taiwan from

Table A.4
Deterrence Success in the Face of a Highly Motivated Aggressor

Case	Aggressor Motivation	U.S. Clarity	Local Balance of Forces	U.S. Interests	Adversary Belief in U.S. Response	Notes[a]
Military Assistance Advisory Group in Republic of China (Taiwan)	High	Clear	U.S. advantage	High	Yes	From April 1951 to December 1978, the U.S. Military Assistance Advisory Group provided arms, training, and advice to the Taiwanese military, and monitored implementation of the 1955 Sino-American Mutual Defense Treaty, which prevented mainland China from invading Taiwan. In 1979 the United States recognized the PRC and the last U.S. military personnel departed Taiwan.
First Taiwan Strait Crisis	High	Clear	U.S. advantage	High	Uncertain	PRC and Taiwanese forces engaged in low-level conflict during the First Taiwan Strait Crisis (1954–1955). The United States supported Taiwan, and in February 1955 the U.S. Navy helped Taiwanese military personnel and civilians evacuate from the Tachen Islands. The Sino-American Mutual Defense Treaty was signed in March, toward the end of the crisis, cementing U.S.-Taiwan relations.
Second Taiwan Strait Crisis	High	Clear	U.S. advantage	High	Uncertain	During a four-week conflict, PRC forces shelled the Kinmen and Matsu Islands in the Taiwan Strait, but did not invade Taiwan. While the PRC is coded as highly motivated, Taiwan was at this point a U.S. ally resulting in very high U.S. interests.

Table A.4—Continued

Case	Aggressor Motivation	U.S. Clarity	Local Balance of Forces	U.S. Interests	Adversary Belief in U.S. Response	Notes[a]
U.S. deterrent posture in South Korea	High	Clear	Ambiguous	High	Yes	A 1953 Mutual Defense Treaty committed the United States to defend South Korea from external aggression. While the United States and South Korea faced a highly motivated adversary and even an ambiguous local balance of forces, U.S. interests and clarity are both coded high. The high level of U.S. force presence (and their trip-wire effect) contributes to North Korean confidence in a U.S. response to territorial aggression.
Korean DMZ Conflict	High	Clear	Ambiguous	High	Yes	This case consisted of a series of low-level clashes between U.S./ROK and DPRK forces along the DMZ from 1966 to 1969. Similar to the general deterrence case, while North Korea is coded as highly motivated and the local balance of forces as ambiguous, U.S. interests, clarity, and trip-wire forces served as a counter.
1980s buildup of U.S. forces in Korea	High	Clear	U.S. advantage	High	Yes	U.S./ROK—DPRK tensions increased throughout the 1980s (e.g., the 1983 Rangoon bombing and the 1987 Korean Air flight 858 bombing). As the Soviets supported North Korea, U.S. force presence on the peninsula increased from roughly 40,000 to 50,000, and the United States sold F-16s to South Korea. Similar to the general deterrence case, while North Korea is coded as highly motivated and the local balance of forces is coded as ambiguous, U.S. interests, clarity, and trip-wire forces served as a counter.

Table A.4—Continued

Case	Aggressor Motivation	U.S. Clarity	Local Balance of Forces	U.S. Interests	Adversary Belief in U.S. Response	Notes[a]
Sino-Indian War of 1962	High	Somewhat clear	Adversary advantage	Medium	Uncertain	During the 1962 Sino-Indian War, the United States sought to deter China—provoked by perceived Indian interference in Tibetan issues and border disagreements—from escalating/expanding the war. The United States provided India with military assistance and deployed the USS *Kitty Hawk* to the Bay of Bengal in an effort to signal support.
U.S. show of force in Panama (Operation Nimrod Dancer)	High	Clear	U.S. advantage	High	Yes	In response to tensions and provocations between Manuel Noriega's Panamanian defense forces and U.S. military forces stationed in the Canal Zone, in May 1989 the United States sent 1,900 troops to bolster the 11,000 U.S. troops already stationed there to defend U.S. civilians and property. Over the summer and fall, U.S. soldiers in the Canal Zone began training for a potential invasion. This case ended with the December 20, 1989, invasion of Panama.
U.S. deterrence posture in the Persian Gulf (Operation Desert Shield)	High	Clear	U.S. advantage	High	Yes	Following Iraq's August 1990 invasion of Kuwait, the United States sought to assure the defense of Saudi Arabia and its oil fields from Iraq. From early August, the United States quickly built up forces in Saudi Arabia. By November, U.S. political objectives had changed from the initial limited aims to the liberation of Kuwait. Operation Desert Storm began on January 17, 1991, ending this case.

Table A.4—Continued

Case	Aggressor Motivation	U.S. Clarity	Local Balance of Forces	U.S. Interests	Adversary Belief in U.S. Response	Notes[a]
1996 Taiwan Strait Crisis	High	Somewhat clear	Ambiguous	High	Uncertain	As cross-strait tensions escalated from 1995–1996, the PRC conducted a series of missile tests, mobilized forces, and undertook a series of naval and amphibious assault exercises. The United States sent a large military task group into the region, including two aircraft carrier groups. While U.S. recognition of the PRC and the end of the Sino-American Mutual Defense Treaty in 1979 meant Taiwan was no longer a U.S. ally, the United States still had vested interests, and the PRC had not attained an advantage in the local balance of forces.

[a] Notes for some cases developed from analysis in Jennifer Kavanagh et al., unpublished RAND research (RUGID deterrence cases).

Figure A.1
Regional Distribution of Deterrence Success in the Face of a Highly Motivated Aggressor

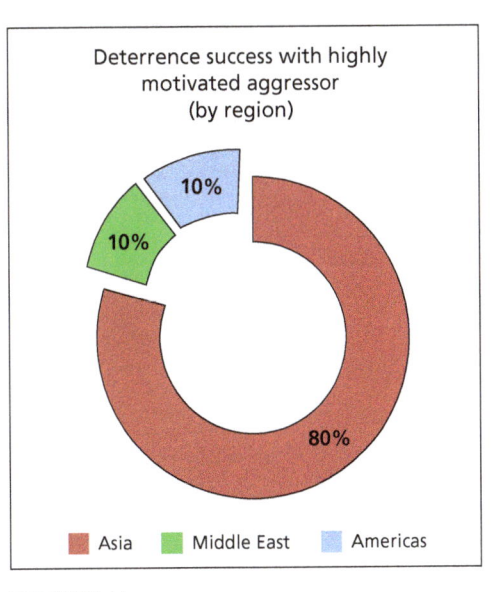
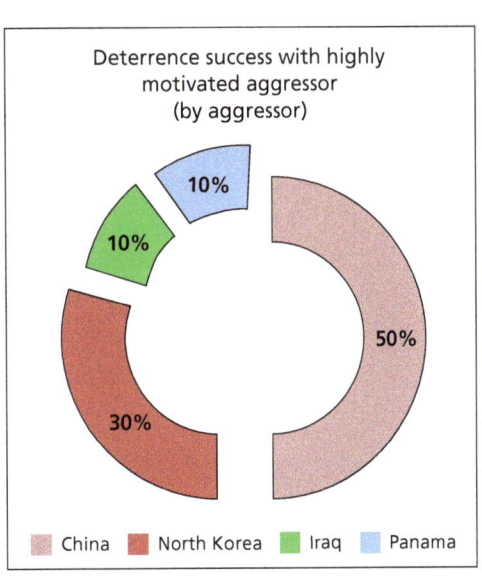

China from 1951 until 1979, and "immediate" deterrence cases for the First, Second, and 1996 Taiwan Straits Crises. Three cases occurred on the Korean Peninsula: the general U.S. deterrence posture in South Korea from 1953 to the present, and two "immediate" deterrence efforts, during the Korean DMZ Conflict from 1964 to 1969 and during the 1980s buildup of U.S. forces in Korea. The United States also sought to deter China from expanding the geographic scope of the 1962 Sino-Indian War. Outside East Asia, the United States successfully deterred two highly motivated aggressors: during the 1989 U.S. show of force in Panama (Operation Nimrod Dancer), when the United States sought to defend the Panama Canal Zone from the Panamanian government, and during Operation Desert Shield, when the United States sought to defend Saudi Arabia from Iraq.

U.S. extended deterrence efforts also succeeded at times when the United States was only coded as somewhat clear or even ambiguous about what it wanted from the adversary and the potential consequences (see Table A.5). For example, the United States was only somewhat clear during efforts to deter Egypt during the Suez Crisis, during the U.S. show of force during the Laos Crisis, while discouraging the expansion of the Sino-Indian War of 1962, and during the 1996 Taiwan Strait Crisis. The United States was ambiguous in only one case of successful extended deterrence, the U.S. Cold War deterrent posture in Iran; however, the aggressor in this case—the Soviet Union—was also coded as having low motivation.

Finally, the United States successfully deterred adversaries four times in the face of a clear adversary advantage in the local balance of forces (see Table A.6). In the

Table A.5
Deterrence Success with Limited or Ambiguous Clarity

Case	Aggressor Motivation	U.S. Clarity	Local Balance of Forces	U.S. Interests	Adversary Belief in U.S. Response	Notes[a]
U.S. Cold War deterrent posture in Iran	Low	Ambiguous	Adversary advantage	Medium	Uncertain	Despite U.S. ambiguity, there is scant evidence of USSR motivation/desire to undertake territorial aggression against Iran.
Suez Crisis (deter counter-attack on Israel)	Medium	Somewhat clear	Ambiguous	Medium	Yes	After the French, Israeli, and UK invasion of the Sinai Peninsula in response to Gamal Abdel Nasser's nationalization of the Suez Canal, the United States implemented sanctions against the USSR and applied political pressure to deter a counterattack against Israel.

Table A.5—Continued

Case	Aggressor Motivation	U.S. Clarity	Local Balance of Forces	U.S. Interests	Adversary Belief in U.S. Response	Notes[a]
Laos Crisis, U.S. show of force (1962)	Medium	Somewhat clear	Ambiguous	High	Uncertain	The United States sought to discourage the North Vietnamese intervention in Laos from spilling over to neighboring Thailand. The United States deployed the Marines, but U.S. force levels never exceeded 3,500, as the situation in Laos began to stabilize in mid-1962.
Sino-Indian War of 1962	High	Somewhat clear	Adversary advantage	Medium	Uncertain	During the 1962 Sino-Indian War over disputed borders at Aksai Chin and along the McMahon Line, the United States sought to deter China from escalating/expanding the war. While the United States was not explicit about whether it would intervene, it did provide India with military assistance and deployed the USS *Kitty Hawk* to the Bay of Bengal in an effort to signal support. And, from the end of the war until the 1964 Chinese development of nuclear weapons, the United States was committed to the territorial defense of India from "communist aggression."
1996 Taiwan Strait Crisis	High	Somewhat clear	Ambiguous	High	Uncertain	As cross-strait tensions escalated in 1995–1996, the PRC conducted a series of missile tests, mobilized forces, and undertook a series of naval and amphibious assault exercises. The United States sent a large military task group into the region, including two aircraft carrier groups. While U.S. recognition of the PRC and end of the Sino-American Mutual Defense Treaty in 1979 meant Taiwan was no longer a U.S. ally, the United States still had vested interests involved and the PRC had not attained an advantage in the local balance of forces.

[a] Notes for some cases developed from analysis in: Jennifer Kavanagh et al., unpublished RAND research (RUGID deterrence cases).

Table A.6
Deterrence Success Where There Was U.S. Disadvantage in the Local Balance of Forces

Case	Aggressor Motivation	U.S. Clarity	Local Balance of Forces	U.S. Interests	Adversary Belief in U.S. Response	Notes[a]
U.S. Cold War deterrent posture in Iran	Low	Ambiguous	Adversary advantage	Medium	Uncertain	Despite a U.S. disadvantage in the local balance of forces and U.S. ambiguity, there is scant evidence of USSR motivation/desire to undertake territorial aggression against Iran.
1948 Berlin Crisis	Medium	Clear	Adversary advantage	High	Uncertain	The Soviet blockade of the Western powers' road, rail, and canal access to West Berlin promoted an airlift to resupply the city. Despite an adversary advantage in the local balance of forces, U.S. interests and clarity were high.
1961 Berlin Crisis	Medium	Clear	Adversary advantage	High	Uncertain	Over several years, the Soviets pushed for NATO military forces to leave West Berlin. The crisis culminated in the October 1961 tank standoff at Checkpoint Charlie. Despite an adversary advantage in the local balance of forces, U.S. interests and clarity were high.
Sino-Indian War of 1962	High	Somewhat clear	Adversary advantage	Medium	Uncertain	During the 1962 Sino-Indian War over disputed borders at Aksai Chin and along the McMahon Line, the United States sought to deter China from escalating/expanding the war. While the United States had a disadvantage in the local balance of forces, it provided India with military assistance, arms, and advisers, and also sent the USS *Kitty Hawk* to the Bay of Bengal. From the end of the war until the 1964 Chinese development of nuclear weapons, the United States was committed to the territorial defense of India from "communist aggression."

[a] Notes for some cases developed from analysis in: Jennifer Kavanagh et al., unpublished RAND research (RUGID deterrence cases).

U.S. Cold War deterrent posture in Iran case, the United States also only communicated threats ambiguously, but the Soviets seemed to accept the status quo and were low in motivation. In the 1948 and 1961 Berlin crises, the United States had a high degree of interest and was clear about what it wanted and what the consequences would be; and while the Soviets were uncertain whether the United States would respond, their degree of motivation was not high.

Conclusion

We coded 39 cases using five variables that capture the elements of the framework developed in Chapter Two: aggressor motivation, U.S. clarity, the local balance of forces, the degree of U.S. interests involved, and adversary belief that the United States would respond. This approach enabled us to code each case using factors that served as proxy variables for each of the main categories in the framework.

This diverse set of U.S. extended deterrence cases since 1945 yielded findings generally consistent with the literature. They showed no deterrence failures when there was low aggressor motivation, when the United States was clear, or when there was a clear U.S. and allied advantage in the local balance of forces. There were a number of outlier cases, such as U.S. deterrence successes in the face of an adversary's advantage in the local balance of forces and U.S. deterrence failures when the aggressor believed the United States would respond. However, in general, the case study analysis supported the literature on when deterrence succeeds and why.

In particular, this analysis confirms nine of the 16 variables outlined in the initial framework of factors governing deterrence success given in Chapter Two. Table A.7 highlights the specific variables emphasized by the quantitative analysis. As this appendix has

Table A.7
Quantitative Analysis: Confirmed Variables

Category	Variable
How intensely motivated is the aggressor?	1. General level of dissatisfaction with status quo and determination to create a new strategic situation.
	2. Degree of fear that the strategic situation is about to turn against the aggressor in decisive ways.
	3. Level of national interest involved in specific territory of concern.
	4. Urgent sense of desperation or requirement to act; whether aggressor is locked into course of action.
	5. Degree of aggressive, reckless, risk-accepting opportunism.
	6. Level of motivated reasoning in play; degree of wishful thinking, misperception of basic strategic context.

Table A.7—Continued

Category	Variable
Is the defender clear and explicit regarding what it seeks to prevent and what actions it will take in response?	1. Precision in the type of aggression the defender seeks to prevent.
	2. Clarity in the actions that will be taken in the event of aggression.
	3. Forceful communication of these messages to outside audiences, especially potential aggressor(s).
	4. Timely response to warning with clarification of interests, threats.
Does the aggressor view the defender's threats as credible and intimidating?	1. Actual and perceived strength of the local military capability to deny the presumed objectives of the aggression.
	2. Degree of automaticity of defender response, including escalation to larger conflict.
	3. Degree of actual and perceived credibility of political commitment to fulfill deterrent threats.
	4. Degree of national interests engaged in state to be protected.
	5. Reputation for resolve with potential aggressor.
	6. Degree of threat posed to aggressor's values and interests by the specific responses threatened by defender.

suggested, these tend to focus on a few key issues: the overall motivation of the aggressor; the national interests involved on both sides; the clarity, precision, and degree of communication of U.S. deterrent threats and statements; relative military strength; and the perceived credibility of deterrent threats, including the aggressor's belief in U.S. willpower.

We now turn to the qualitative case studies to determine their lessons for the factors affecting success in extended deterrence.

APPENDIX B
Qualitative Case Study Analyses: Berlin

Despite varied incentives to remove the Western presence from Berlin and a clear ability to overwhelm the limited defenders in the city, the Soviet Union chose not to do so throughout the entire duration of the Cold War. In the main, West Berlin therefore serves as an example of successful U.S. and allied deterrence. While conflict was avoided, however, the fact that there were multiple acute crises over the city that could have easily led to war should temper our assessment of the success of U.S. policies, particularly in the earlier periods. This appendix will briefly review the history and key strategic developments surrounding Berlin during the Cold War.

1945–1957

The end of the Second World War left Germany divided into separate zones of occupation among the victorious allies. This division into the American, British, French, and Soviet Zones was mirrored in the capital of Berlin, which was otherwise isolated in the middle of the Soviet Occupation Zone in East Germany. Agreements made at the Yalta Conference in 1945 had provided for the creation of these zones and stated that affairs in Germany should be run on the principle of consensus among the four powers, but had left many other vital details unspecified.[1] Most notable among these omissions were the future governance of Germany itself and, crucially, the right of Western ground access to the American, British, and French Zones in Berlin.[2] These ad hoc arrangements were placed under considerable strain as relations among the former allies deteriorated throughout the late 1940s. The U.S. proposal, and Soviet rejection, of the Marshall Plan in 1947, the Greek Civil War and President Harry S. Truman's announced policy of containment, and the Soviet execution of a coup in

---

[1] Thomas Parrish, *Berlin in the Balance 1945–1949: The Blockade, the Airlift, the First Major Battle of the Cold War*, Reading, Mass.: Addison-Wesley, 1998, pp. 24–25; Daniel F. Harrington, *Berlin on the Brink: The Blockade, the Airlift, and the Early Cold War*, Lexington: University Press of Kentucky, 2012, pp. 7, 15.

[2] Air access rights had been specified in a 1945 agreement, the greater difficulties of air traffic control having required greater early codification of collaboration in this area. See Parrish, 1998, pp. 134–135.

Czechoslovakia in 1948 all reflected the divisions that were becoming starkly drawn between the former Allies in Europe.[3]

While this decline in East-West relations and the development of the Cold War had many causes, the proximate issue of greatest concern in central Europe was the nature of the future German government. The reorganization of postwar Europe took on substantial urgency with the failure of European economies to quickly recover in the aftermath of the war, and concerns that the lack of economic stability could lead to political instability as well.[4] The United States had proposed a massive cooperative program of economic reconstruction in Europe in 1947 (what would become the Marshall Plan). Although the Soviets and their satellite states had initially been invited to participate, they refused. The German economy, previously the strongest in Europe, was considered vital to the success of a European recovery, and the American and British Zones in West Germany began to integrate economically, including through the introduction of a new German currency, forming the basis for a future West German state.[5]

By contrast, Soviet plans for Germany continued to call for a unified state involving all four zones, albeit one with borders of its choosing (much of what had previously been eastern Germany having been ceded to Poland to compensate for the large sections of previous Polish territory ceded to the Soviet Union in 1945), and a government dominated by socialist or communist factions and open to Soviet influence.[6] Moscow strongly opposed Western plans to create a separate West Germany and the specter they raised of a revitalized, possibly rearmed German state closely allied with Britain, France, and the United States.

In order to try to forestall the development of a West Germany, Joseph Stalin decided to use the leverage he had over the exposed Western zones in Berlin. In March 1948 the Soviet Union imposed a blockade of ground and rail traffic into the Western sectors of Berlin on the grounds that with the development of a West German state, the USSR needed to protect its zone from "economic disruption."[7] This forced the Western Allies in the near term to begin relying on emergency stockpiles of food and coal, and in the long term it presented them with a number of unappealing choices: to allow the citizens of West Berlin slowly to starve, to withdraw Western forces from Berlin and cede the city in full to the Soviets, to accede to Soviet demands to stall or reverse the development of West Germany, or to test the ground blockade with force, which both sides felt could risk open war. A final option, to fully supply the city indefinitely via the air and leakages in the ground blockade, was not initially

---

[3] John Lewis Gaddis, *The Cold War: A New History*, New York: Penguin Books, 2006, pp. 31–33.

[4] Avi Shlaim, *The United States and the Berlin Blockade, 1948–1949: A Study in Crisis Decision-Making*, Berkeley: University of California Press, 1983, p. 29.

[5] Shlaim, 1983, p. 30.

[6] Shlaim, 1983, pp. 22–23.

[7] Parrish, 1998, p. 234.

considered to be viable.[8] By threatening the Western position in Berlin in this manner, the Soviets hoped either to force the West to reconsider its plans for the future of Germany, or to force it to withdraw from Berlin in humiliating fashion, calling into question Western unity and American willingness to remain engaged in Europe and defend its wartime allies.[9] In the end, the Soviets achieved neither objective, in large part due to the unexpected success of the West in supplying the city through a protracted airlift. This effort bought the West time, provided a clear mechanism to demonstrate its resolve and commitment to defending its position in Berlin, and hardened political attitudes in Germany against the Soviet Union.

Stalin eventually withdrew his conditions, and the Soviet blockade, as well as the Western counterblockade imposed in retaliation, was lifted in May 1949.[10] The principle of Western access to Berlin and U.S. commitment to remaining in the city thereby demonstrated, the status quo persisted throughout the early to mid-1950s.

1958–1964

The arrangement between the East and the West in Berlin continued unchanged throughout the remainder of the Truman administration and much of that of his successor, President Dwight D. Eisenhower. However, both economic and strategic developments combined to undermine Soviet support for this arrangement by the end of the 1950s, setting the stage for an additional crisis from 1958 through 1961.

The fusion of the American, British, and French Zones into an independent West Germany in 1949 led to a robust economic recovery and a stable, politically vibrant state (by comparison) based in Bonn. Throughout the 1950s, economic growth in West Germany averaged nearly 8 percent per year.[11] The less effective economic recovery in the East, worsened by Soviet appropriation of East German capital and infrastructure after the war to pay reparations and the transition to a socialist system, combined with the draconian political restrictions in the new East Germany, created a large incentive for Germans—particularly younger, skilled workers—to migrate from the East to the West.[12] A partitioned Berlin created the perfect opportunity for them to do so.

Postwar arrangements between the Allies in Berlin had allowed for the free movement of people throughout the city, regardless of the zone in which they resided. While the border between East and West Germany proper was more easily controlled, the

---

[8] Shlaim, 1983, p. 111; Harrington, 2012, pp. 3–4.

[9] Parrish, 1998, pp. 142–143.

[10] Parrish, 1998, pp. 320–321.

[11] Wendy Carlin, "West German Growth and Institutions, 1945–1990," in Nicholas Crafts and Gianni Toniolo, eds., *Economic Growth in Europe Since 1945*, Cambridge: Cambridge University Press, 1996, p. 4.

[12] André Steiner, "From Soviet Occupation Zone to 'New Eastern State,'" in Hartmut Berghoff and Uta Andrea Balbier, eds., *The East German Economy, 1945–2010: Falling Behind or Catching Up?* Cambridge: Cambridge University Press, 2013, pp. 25–26.

arrangements governing West Berlin allowed East Germans to use the city as a means of migrating west. Throughout the 1950s, East Germany's population declined roughly 15 percent, with much of the loss concentrated in younger, higher-skilled labor, calling into question the future viability of the East German state.[13] The Soviets (including new premier Nikita Khrushchev) and their East German client state, led by Walter Ulbricht, were strongly motivated to come up with a solution.

The strategic situation between the East and the West had also evolved substantially since the 1948 crisis. The Western allies had joined together in NATO starting in 1949, while the Soviets formed the corresponding Warsaw Pact among its satellites in 1955. The West had also revitalized its conventional forces in Europe to a greater degree than the late 1940s, although the balance of such forces in Europe by the end of the 1950s was still strongly tilted in favor of the Soviet bloc.[14]

Perhaps most notably, however, the Soviet Union had developed its own nuclear deterrent, including some limited potential to strike the United States directly using intercontinental ballistic missiles (ICBMs).[15] Soviet nuclear forces did continue to lag behind those of the United States, however, and in 1961 U.S. decisionmakers assessed that there was still a possibility of winning a first strike nuclear engagement.[16] They were aware, however, that given the development of Soviet forces this "window of strategic advantage was closing."[17]

From the Soviet perspective, while U.S. nuclear superiority had been a constant concern since the 1940s, the creation of independent European nuclear forces was a further worrying development. The British deployment of independent nuclear capabilities after 1953, and "nuclear sharing" arrangements, wherein the United States placed nuclear weapons in NATO member countries functionally under host government control after the late 1950s, increased Soviet concerns.[18] However, it was the possibility of a rearmed West Germany with a fully independent nuclear force that prompted the greatest anxiety in Moscow, given the recent history of German-Soviet

---

[13] Steiner, 2013, p. 26.

[14] At the time, Western estimates of the disparity of forces ranged between five to one and ten to one in favor of the Warsaw Pact, although it is worth noting that these estimates, although believed in the West, likely overstated Eastern Bloc capabilities in reality. The Soviets, presumably, had a more accurate sense of their own capabilities, which still likely gave them a substantial advantage. See Richard A. Bitzinger, *Assessing the Conventional Balance in Europe, 1945–1975*, Santa Monica, Calif.: RAND Corporation, N-2859-FF/RC, 1989, pp. 7–10.

[15] Richard L. Kugler, *The Great Strategy Debate: NATO's Evolution in the 1960s*, Santa Monica, Calif.: RAND Corporation, N-3252-FF/RC, 1991, p. 9.

[16] Marc Trachtenberg, *A Constructed Peace: The Making of the European Settlement, 1945–1963*, Princeton, N.J.: Princeton University Press, 1999, pp. 295–296. The USSR lagged particularly in delivery systems such as strategic bombers and ICBMs, where in 1961 the United States enjoyed roughly a ten-to-one advantage. See Kugler, 1991, p. 9.

[17] Trachtenberg, 1999, p. 296.

[18] Trachtenberg, 1999, pp. 194–195. Although the United States theoretically maintained control over these shared weapons, in practice this control appears to have often been ineffective or insufficient.

conflict and the West German refusal to accept a separate East German state under Soviet influence.[19]

The second crisis over Berlin began in November 1958 when Khrushchev announced that the Soviet Union would conclude a separate peace treaty with the East German state unless the West agreed to relinquish its position in West Berlin.[20] Such a treaty could have liquidated the legal basis of the Western Zones in Berlin, signaling a Soviet policy of treating them as illegitimate and perhaps indicating a willingness to use force to eject the West from the city.[21] While the Eisenhower administration was willing to discuss the matter with the Soviets, in part because it recognized that over the long run the Western position in Berlin was simply not tenable without Soviet acquiescence, its willingness to compromise did not include any steps that might have called into question the administration's commitment to collective security and the defense of Europe, which since 1948 had been clearly established to include West Berlin.[22] With U.S. elections on the horizon, Khrushchev hedged his bets that an incoming administration of John F. Kennedy might prove more pliable, even going so far as to refuse to release captured U.S. pilots until after the election in a bid to damage the ardently anticommunist Richard Nixon and to boost Kennedy.[23]

The new Democratic Party–led administration did not improve relations as the Soviets expected, however, as a series of early diplomatic overtures were misinterpreted;[24] by summer 1961, Khrushchev was repeating his insistence that he intended to negotiate a separate peace treaty with East Germany and liquidate the legal basis for the Western occupation zones, stating that West Berlin was in reality East German territory.[25] The U.S. administration's response was initially muted, but by July 25 Kennedy gave a clear, televised address indicating that the United States viewed its position in Berlin as inviolate, and an issue over which it was willing to go to war.[26] Kennedy also

---

[19] Trachtenberg, 1999, p. 252.

[20] Trachtenberg, 1991, p. 251.

[21] Trachtenberg, 1999, p. 251.

[22] Trachtenberg, 1999, pp. 258–261.

[23] Frederick Kempe, *Berlin 1961: Kennedy, Khrushchev, and the Most Dangerous Place on Earth*, New York: G. P. Putnam's Sons, 2011, pp. 38–39, 73–75.

[24] Khrushchev initiated a series of diplomatic overtures to Kennedy coinciding with the latter's inauguration, including the release of captured U.S. pilots, a reduction in censorship of Western news sources, and increased student exchanges. At the same time, however, Kennedy was briefed on the contents of a speech Khrushchev had given to Soviet officials the previous November that included fiery (albeit fairly standard) anti-Western language. Kennedy interpreted the speech as the true evidence of Soviet intentions, and declined to reciprocate Khrushchev's overtures. See William Taubman, *Khrushchev: The Man and His Era*, New York: W. W. Norton and Company, 2003, pp. 487–488; and Kempe, 2011, pp. 75–77.

[25] Kempe, 2011, pp. 77–78, 245–246.

[26] W. R. Smyser, *Kennedy and the Berlin Wall*, Lanham, Md.: Rowman and Littlefield Publishers, 2009, p. 88; Kempe, 2011, p. 314; Richard D. Williamson, *First Steps Toward Détente: American Diplomacy in the Berlin Crisis, 1958–1963*, Lanham, Md.: Lexington Books, 2012, p. 98.

signaled, however, that what the Soviets did in East Berlin was not something over which the United States was willing to fight.[27]

Khrushchev took this signal and acted accordingly to solve the immediate migration problem. In violation of postwar agreements on Berlin, the Soviets and East Germans rapidly constructed a barrier separating East and West Berlin in the early morning of August 13, 1961.[28] Over time the barrier was gradually converted into a permanent, and more formidable, wall. Western responses to the new Berlin Wall were initially muddled. While it was within Western rights to tear the barrier down immediately, and despite Soviet fears that the West would do so, the construction of the wall was allowed to proceed unimpeded.[29] However, local commanders on the ground, including GEN Lucius D. Clay, who ran West Berlin during the 1948 Berlin Airlift and was recalled to the city out of retirement, did engage in a series of probes and provocations of the new arrangement without authorization from Washington, including practicing maneuvers to bulldoze the wall. These maneuvers were observed by Moscow, but not reported to Washington.[30] Authorized Western efforts focused on maintaining Allied military rights to transit the city without restriction, and initially on refusing to acknowledge any East German role in the Soviet Zone. The exercise of these rights, including a refusal to acknowledge East German—rather than Soviet—authority in Berlin, together with a Soviet assessment that the United States was preparing an operation to tear down the wall timed to humiliate Khrushchev at a party congress in October, led to a tense tank standoff on the night of October 27, 1961.[31]

In the end, however, the United States—through both formal talks and a back-channel negotiation between Attorney General Robert F. Kennedy and a Soviet spy in New York—signaled its willingness to accept the Soviet construction of the wall and restrictions on civilian travel as long as the Soviets continued to respect Allied military access to and position in the city, and provided that Soviet tanks withdrew from the intra-Berlin border first.[32] The Soviets complied, and the standoff ended the next day.

Tensions over the city recurred in 1962, both before and during the Cuban Missile Crisis, when Kennedy became concerned that Soviet bases in Cuba were being established to provide the Soviet Union leverage to dislodge the West from Berlin.[33]

---

[27] Smyser, 2009, p. 88; Kempe, 2011, p. 359.

[28] Smyser, 2009, p. 101; Kempe, 2011, p. 352.

[29] Smyser, 2009, p. 106; Kempe, 2011, pp. 359–360.

[30] Kempe, 2011, p. 418.

[31] Kempe, 2011, pp. 448–449, 468.

[32] Smyser, 2009, p. 142; Kempe, 2011, pp. 479–481.

[33] Kempe, 2011, p. 495.

In continuing negotiations over Berlin in fall 1962, the Soviets again threatened to sign a separate peace treaty with East Germany and dislodge the West from Berlin—by force if necessary.[34] However, after Kennedy's October 22 speech stating that any action against Berlin during the crisis would be met with force, Khrushchev removed troops from the border with West Germany to signal that he was not seeking a horizontal escalation of tensions.[35] After the resolution of the Cuban Missile Crisis, the United States and the Soviet Union by 1963 had reached a tacit agreement to accept the status quo over Berlin and Germany more broadly, an agreement implicitly linked to nuclear arms control agreements such as the Limited Nuclear Test Ban Treaty and an understanding that West Germany would remain a nonnuclear power.[36] After 1963 Berlin did not again become the site of an acute superpower crisis for the duration of the Cold War.

1964–1989

Relations between the East and the West over Germany in general and Berlin in particular stabilized after the mid-1960s. By the early 1970s, West German politicians adopted a new policy of engagement and rapprochement (*Ostpolitik*) with East Germany and the Soviet Union that paved the way for formal diplomatic recognition of the German Democratic Republic.[37] This shift contributed to an overall policy of détente between the East and the West, culminating in the Helsinki Accords in 1975, in which the two sides pledged mutual recognition of the status quo territorial arrangement in Europe.[38]

The Four Power Agreement on Berlin, fully codifying the arrangements governing the city, including access rights and the legal status of the occupation zones, was signed in 1971.[39] The rules tacitly agreed to in the aftermath of the construction of the Berlin Wall—continued Western military access and transit rights and engagement with East German rather than Soviet authorities—continued to govern operations in Berlin until the dissolution of the wall in 1989.

The foundation for this relative stability was the strategic arrangement worked out between the United States and the Soviet Union during the 1958–1962 crisis period: the Soviets would continue to permit Western military presence in and access to their zones in West Berlin, while the West would permit the Soviets to manage East Berlin

---

[34] Trachtenberg, 1999, pp. 349–350.

[35] Kempe, 2011, p. 497.

[36] Trachtenberg, 1999, pp. 382, 386–390.

[37] Tony Judt, *Postwar: A History of Europe Since 1945*, New York: Penguin Press, 2005, pp. 497–498.

[38] Judt, 2005, p. 501.

[39] Gunther Doeker, Klaus Melsheimer, and Dieter Schroder, "Berlin and the Quadripartite Agreement of 1971," *American Journal of International Law*, Vol. 67, No. 1, January 1973, p. 54.

however they saw fit.[40] More broadly, the Soviets would accept the Western presence in Berlin as long as West Germany remained a nonnuclear state—an arrangement that, to be accepted by Bonn, would in turn require a substantial and indefinite American military presence in West Germany.[41] This arrangement survived several other notable developments in the strategic balance between the two sides in Europe throughout the remainder of the Cold War.

From the early 1960s onward, the United States gradually moved to increase its military capabilities in West Germany to allow for a more flexible response to a Soviet provocation or attack rather than a more immediate need to escalate to nuclear weapons or face conventional military defeat. The difference from the previous U.S. posture, however, was more one of degree than kind.[42] Already in 1960 the United States had roughly 250,000 troops in West Germany.[43] This number actually declined somewhat during the Vietnam War in the late 1960s before rebounding, and even modestly increasing, during the U.S. military buildup of the 1980s.[44] The increase in conventional forces along with the adoption of flexible response was therefore less notable than the increase that had already occurred from 1950 to 1960 (when troop numbers more than doubled), a period during which U.S. policy under Eisenhower continued to meet any Soviet attack across the intra-German border with nuclear retaliation.[45] In the 1960s, U.S. forces in Europe did shift to become much more heavily armored and mechanized, increasing their combat capabilities.[46] In the main, however, the shift to flexible response was as much a shift in U.S. doctrine and signaling as it was a shift in U.S. capabilities, at least in West Germany.

While the size of U.S. forces in West Germany did not increase notably throughout this period, the United States did take additional steps to signal that it could flow additional forces from the United States quickly in a crisis. To reassure nervous NATO allies over the withdrawal of forces from Europe during the Vietnam War, in 1969 the United States held the first large-scale Return of Forces to Germany (REFORGER) exercise, consisting of roughly 12,000 troops and using pre-positioned equipment.[47] While initially only demonstrating the ability to quickly return forces recently with-

---

[40] Kempe, 2011, pp. 489–491.

[41] Trachtenberg, 1999, pp. 398–400.

[42] Trachtenberg, 1999, pp. 288–289.

[43] Defense Manpower Data Center, *Historical Report—Military Only* (aggregated data 1950–current), Alexandria, Va.: U.S. Department of Defense, 2017.

[44] Defense Manpower Data Center, 2017.

[45] Trachtenberg, 1999, pp. 288–289; Defense Manpower Data Center, 2017.

[46] Kugler, 1991, p. 104.

[47] Kugler, 1991, pp. 103–104; U.S. Army, "Countdown to 75: US Army Europe and REFORGER," March 22, 2017.

drawn, REFORGER grew in size throughout the later Cold War period, until by 1988 the costly exercises included roughly 75,000 U.S. troops and thousands of tanks and armored personnel carriers and strongly indicated a U.S. commitment and ability to rapidly flow forces to Europe in the event of a crisis.[48] The Soviets took these large-scale exercises seriously; they put forward numerous proposals throughout the 1970s and 1980s to limit the size and frequency of military exercises in Europe.[49]

Two notable strategic changes did occur in this period that related to nuclear forces. First, by the mid- to late 1960s, the Soviet Union had achieved rough parity with the United States in its nuclear delivery systems.[50] The two sides engaged in a massive arms race throughout the Cold War period, building tens of thousands of nuclear weapons, but neither side achieved any clear strategic advantage over the other, as the Americans had enjoyed in the early Cold War period.

The second change had to do with the development and deployment of intermediate-range missiles. Beginning in the late 1970s, the Soviet Union began to deploy intermediate-range nuclear-armed missiles in Europe. NATO allies became concerned that these gave the Soviets the ability to rapidly strike targets in Western Europe, while a NATO response would ultimately depend on a United States decision to launch ICBMs from the United States itself, presumably putting the United States at greater risk in a retaliatory strike. This geographic asymmetry gave rise to fears that, faced with only minutes to respond after a Soviet launch, the United States might hesitate to put its own territory at risk, and that this perception itself could encourage Soviet nuclear threats and brinksmanship. This prompted the United States to develop and deploy intermediate-range missiles such as the Pershing II to Europe starting in 1983, over strong popular protests. While the Soviets initially reacted negatively, the U.S. missile deployment was eventually a key factor in motivating the Soviets to conclude the Intermediate Range Nuclear Forces Treaty in 1987, which eliminated the missiles altogether. While this episode raised concerns over strategic stability in Europe, and arguably called into question the overall tacit agreement under which the two sides had operated in Europe since 1963, Berlin was not singled out as an area of competition, as it had been in earlier crises. Rather, the

---

[48] Michael R. Gordon, "U.S. War Game in West Germany to Be Cut Back," *New York Times,* December 14, 1989.

[49] In part, Soviet initiatives to restrict large-scale exercises (which NATO resisted) reflected an attempt to disadvantage NATO's defense strategy in Europe, and this required a greater emphasis on reinforcement, and thus a greater utility for exercises to practice such reinforcement. Large-scale NATO exercises were also sometimes seen as threatening by the Soviets, as evidenced by the events surrounding the Able Archer exercise in 1983. See Robert D. Blackwill and Jeffrey W. Legro. "Constraining Ground Force Exercises of NATO and the Warsaw Pact," *International Security,* Vol. 14, No. 3, Winter 1989–1990, pp. 86–88; and Dmitry Dima Adamsky, "The 1983 Nuclear Crisis–Lessons for Deterrence Theory and Practice," *Journal of Strategic Studies,* Vol. 36, No. 1, 2013, pp. 12, 22–28.

[50] Kugler, 1991, pp. 9–10.

U.S. and European concern was for the protection of not just Berlin but also Frankfurt, London, and Paris.

The Western position in Berlin was therefore maintained, against initial expectations, throughout the Cold War, until the collapse of the Berlin Wall in November 1989 and the reunification of Germany in early 1990. The following section will assess how that came to pass.

Variables Affecting Deterrence Failure and Success

1945–1957

How Motivated Was the Soviet Union to Eject the Western Allies from Berlin?
Throughout the Berlin crisis that began in 1948, the Soviet Union adopted two somewhat contradictory goals. The first Soviet goal in threatening the Western allies' position in West Berlin was to change Allied policy regarding the establishment of an independent West Germany.[51] West Berlin was a means to that end. The Soviets strongly opposed the creation of a West Germany that was capitalist, allied with the West, and with the potential to rearm; and they were highly motivated to use what leverage they had to prevent this development. However, pressure on the Allied position in Berlin was not purely instrumental. The second Soviet goal was indeed to try to eject the Allies from Berlin, or at least severely restrict their ability to independently operate their zones.[52] In the event that the creation of West Germany proved unstoppable, then removing the Western presence from Berlin, or at a minimum ensuring that no West German currency would be used, would have been strongly preferable to the alternative of allowing the Allies to operate their half of the city as they saw fit in a sensitive enclave deep inside the Soviet Zone in East Germany.[53] The more the Soviets were successful in forcing the Allies out or restricting their independence before a resolution of the status of West Germany, however, the more it would have eliminated the Soviets' potential leverage.[54] The tension between these goals was never truly resolved in the Soviet approach to the crisis.

Soviet motivations to apply pressure to West Berlin did not extend to a desire to take the enclave by force. Soviet conventional forces in Germany at this time vastly outnumbered those of the West, and a conventional military solution in Berlin was, in

---

[51] Parrish, 1998, pp. 142–143; Harrington, 2012, pp. 44–45.

[52] Harrington, 2012, p. 44; Michail M. Narinskii, "The Soviet Union and the Berlin Crisis," in Francesca Gori and Silvio Pons, eds., *The Soviet Union and Europe in the Cold War, 1943–1953*, New York: St. Martin's Press, 1996, pp. 65, 69.

[53] Parrish, 1998, pp. 142–144.

[54] Harrington, 2012, pp. 44–45.

principle, achievable.⁵⁵ Despite this capability, a Soviet plan to invade West Berlin was never strongly considered, publicly suggested, practiced, or messaged at this time.⁵⁶ A number of factors contributed to the Soviet reticence, including the U.S. monopoly on deliverable nuclear weapons, the potential costs of such a war and occupation, and general war-weariness.⁵⁷ The Soviets may have been prepared to militarily enforce the ground blockade. The prospects of direct conflict if the West attempted to challenge the blockade directly worried both sides.⁵⁸ However, at the time the blockade was implemented, the Soviets felt that, at least in Berlin, time was on their side, as the West would not be able to maintain its position indefinitely. The unexpected success of the Western airlift, even during the more difficult winter months, revealed that this assessment was mistaken, and the Soviets agreed to lift the blockade shortly thereafter, despite not having achieved either any pause in the formation of West Germany or the removal of the Western presence from Berlin.

Was the U.S. Deterrent Message Clear and Explicit?

The United States had a substantial internal debate in 1948 regarding whether it could and would stay in West Berlin in the face of the Soviet blockade.⁵⁹ Initially the airlift was not considered likely to be able to sustain the city over the long term. While the Western garrison itself could have been provided for indefinitely, the needs of the city's roughly one million civilians represented a greater challenge.⁶⁰ Some officials, including Secretary of the Army Kenneth C. Royall, favored withdrawing from Berlin.⁶¹ Nonetheless, President Truman privately said in June 1948, "We are going to stay, period," and U.S. diplomats delivered the same message to Moscow, including that the United States would resist "any further act of aggression."⁶²

It is important to clarify precisely what actions the United States was attempting to deter. The United States appears to have made no attempt to suggest that it would take military action in response to the Soviet blockade. The blockade had not been anticipated far in advance, but even once it was in place the United States did not

---

⁵⁵ Donald P. Steury, ed., *On the Front Lines of the Cold War: Documents on the Intelligence War in Berlin, 1946 to 1961*, Washington, D.C.: Center for the Study of Intelligence, 1999; Central Intelligence Agency, *Possibility of Direct Soviet Military Action During 1948*, Washington, D.C.: Central Intelligence Agency, March 30, 1948, pp. 142–146.

⁵⁶ Roger G. Miller, *To Save a City: The Berlin Airlift, 1948–1949*, Washington, D.C.: Air Force History and Museums Program, 1998, p. 25; Parrish, 1998, pp. 143–144; Harrington, 2012, pp. 295–296.

⁵⁷ Steury, 1999, pp. 142–146.

⁵⁸ Parrish, 1998, pp. 176–177; Harrington, 2012, pp. 105–106, 270, 273.

⁵⁹ Parrish, 1998, pp. 136, 183.

⁶⁰ Miller, 1998, p. 19.

⁶¹ Parrish, 1998, p. 182; Harrington, 2012, pp. 83–85.

⁶² Parrish, 1998, pp. 183–185; Harrington, 2012, p. 133.

threaten to use force if it were not removed. Instead, the imposition of the blockade led to a frantic search to identify some previous agreement between the United States and the Soviet Union that formally recognized the right of ground access to the Western zones in Berlin, to no avail.[63]

The United States was clearer in its messaging that it was prepared to respond militarily if the Soviet Union attempted to take West Berlin by force. In addition to the direct diplomatic messages delivered to this effect, the United States pointedly deployed B-29 bombers, the planes used to carry the atomic weapons that were dropped on Japan in 1945, to bases in the United Kingdom in response to the crisis.[64] U.S. forces in West Berlin had also been consistent in enforcing their rights on the ground and preventing individual Soviet troops from acting within Western sectors, by force if necessary. While U.S. messaging regarding defending its rights and troops in West Berlin seems to have been relatively clear, there was substantial ambiguity regarding how the United States would respond if it chose to send a land convoy to West Berlin, which would then be stopped by the Soviets. The United States did initially probe a train crossing, with armed guards under orders only to fire if fired upon, only to have the Soviets switch the train onto a side track to prevent its progress.[65] The train eventually withdrew. The United States was, at least at that time, not willing to signal that it would use force to restore ground access to the city. Throughout the crisis, each side was uncertain about the other's response if a skirmish were to erupt due to Soviet enforcement of the blockade, and worried that miscalculations in this event could lead to war.[66]

Was the U.S. Deterrent Message Credible and Convincing?

The geography and balance of conventional forces in West Berlin suggested that the West would struggle to maintain its position. West Berlin was an enclave inside the larger Soviet Zone in eastern Germany. The United States had only about 2,500 troops in West Berlin in 1948, part of a 6,000-troop Western contingent.[67] By contrast, the Soviets had roughly 90,000 troops in the immediate vicinity, with the ability to reinforce more rapidly and robustly than the Americans, whose forces remained at a very low state of combat readiness.[68] The U.S. Army had approximately 90,000 troops in all of Germany, in comparison with between 500,000 to 1,000,000 Soviet

---

[63] Marshall Zhukov had verbally agreed to these access rights in 1945, but no signed document accompanied those assurances. See Parrish, 1998, p. 28; and Harrington, 2012, p. 20.

[64] The bombers did not actually carry nuclear weapons in this case, and Soviet intelligence was aware of this thanks to a British spy, Donald Maclean. See Miller, 1998, pp. 24–25; and Harrington, 2012, p. 122.

[65] Parrish, 1998, pp. 135–136.

[66] Harrington, 2012, p. 246.

[67] Parrish, 1998, p. 175.

[68] Parrish, 1998, pp. 138, 175.

troops.[69] The American ability to promise credibly to stay in West Berlin, and the apparent Soviet disinterest in pursuing a military solution to the crisis, therefore likely rested substantially on U.S. ability and willingness to deliver nuclear weapons in the event of a conflict, something the United States had demonstrated just three years earlier in Japan.[70] It is worth noting, however, that U.S. policymakers were dissatisfied with having to choose between escalation to nuclear war and capitulation; U.S. commitment to nuclear escalation to reverse a conventional defeat may not have been assured.[71] The Soviets certainly took the possibility seriously, though precisely how seriously was not tested at this time given the lack of Soviet interest in fighting a war over Berlin. Enhancing the credibility of U.S. statements that it would not be pushed out of Berlin, the United States would also have suffered tremendous political and diplomatic damage had it been forced to evacuate Berlin in the face of the blockade, particularly from its Western European allies.[72]

Neither side initially considered it possible for the United States to stay in Berlin indefinitely in the face of the blockade without fighting to break it. Both the Americans and the Soviets thought that alternative means of providing food and supplies to the city would not prove sustainable. The United States, in a sense, stalled in its initial refusal in summer and fall 1948 to make a choice between acceding to Soviet demands or leaving the city. Only the later development of the capability to execute and sustain the airlift at a sufficient size made it possible for the United States to avoid the choice between acquiescing to Soviet demands and starting a war. This capability then shifted the burden of choice to the Soviets, who declined to risk war by further escalating the crisis and instead themselves acquiesced to the creation of a West Germany allied with the United States.[73]

1958–1963

How Motivated Was the Soviet Union to Eject the Western Allies from Berlin?

Soviet goals in the second Berlin crisis again did not focus on the ejection of the Allies from West Berlin. Instead, they focused on other areas linked to West Berlin: the flow of East German immigrants through the city to the West, and the possible leverage the exposed Western position in the city could provide in limiting the militarization, and ultimately nuclearization, of West Germany.[74]

---

[69] Miller, 1998, pp. 16–18. The deployment of B-29s to the United Kingdom during the crisis, even though the Soviets knew they did not carry nuclear weapons, still sent a message underlining U.S. ability to use these weapons. See Harrington, 2012, p. 122.

[70] Parrish, 1998, pp. 138–140.

[71] Parrish, 1998, p. 139.

[72] Parrish, 1998, pp. 189–190.

[73] Narinskii, 1996, pp. 72–73; Harrington, 2012, pp. 270, 273.

[74] Trachtenberg, 1999, p. 252; Kempe, 2011, p. 97.

Soviet motivation to apply pressure to the United States—and to a lesser extent France and the United Kingdom—over Berlin was strong, but throughout the crisis the Soviets remained concerned about the potential for escalation and were not willing to fight a war with the West to achieve their goals.[75] Nonetheless, these goals were important enough to Khrushchev to merit running some risks. The exodus of younger, skilled East Germans had grown acute enough by the late 1950s to threaten the economic viability of the East German state and with it Soviet plans for the development of all of Eastern Europe. Similarly, Soviet fears that a remilitarized, nuclear-armed West German state could prompt or intervene in a popular uprising in East Germany and force Moscow to choose between the loss of their satellite and a nuclear conflict became acute.[76] To address these concerns, Khrushchev was willing to threaten the West with war over Berlin, even if he did not ultimately believe that war would be necessary or likely.[77] Once the United States acquiesced to certain key Soviet goals, such as the building of the wall to limit emigration and the nonnuclear status of West Germany, Soviet motivation to pursue the crisis further was greatly reduced.

Was the U.S. Deterrent Message Clear and Explicit?

Building on the experience of 1948, the United States clearly messaged its determination to stay in West Berlin and to maintain its agreed upon access routes to the city. Through the Eisenhower and Kennedy administrations, the United States never signaled nor suggested the possibility of Western retreat from Berlin, and repeatedly emphasized the maintenance of the status quo regarding West Berlin.[78]

U.S. messaging regarding what it would be willing to accept on other issues, however, varied. With regard to East Berlin and transit rights within the city, the United States at different times sent both conciliatory and aggressive signals. Kennedy, in both direct conversations with Khrushchev in Vienna in July 1961 and through covert back-channel negotiations later that year, was relatively accommodating regarding what steps the Soviets should be permitted to take in their own zone of Berlin.[79] However, due to a misreading of, among other things, unauthorized preparations by local U.S. forces in West Berlin, the Soviets became convinced that the United States planned to use force to tear down the wall at a politically sensitive time for Khrushchev—during the October party congress—making the Soviets more inclined to escalate the crisis through a show of force in response to U.S. probes

---

[75] Taubman, 2003, pp. 505–506; Williamson, 2012, p. 20.

[76] Trachtenberg, 1999, p. 252.

[77] Kempe, 2011, p. 25.

[78] Norman Gelb, *The Berlin Wall: Kennedy, Khrushchev, and a Showdown in the Heart of Europe*, New York: Times Books, 1986, p. 184; Trachtenberg, 1999, p. 261; Kempe, 2011, p. 244; Williamson, 2012, pp. 96–99.

[79] Gelb 1986, pp. 184–185; Kempe, 2011, pp. 247, 479.

regarding transit rights.[80] While the United States was therefore relatively clear about its own red lines, it was less clear that it planned to respect those of the Soviets, and this lack of clarity contributed to the escalation of tensions through fall 1961.

As for the other set of Soviet goals that motivated crisis escalation—Soviet efforts to arrest the military development of West Germany—U.S. positions changed over time. While even during the beginnings of the crisis in 1958 the Eisenhower administration remained supportive of West German efforts to pursue an independent nuclear capability, by 1961 the Kennedy administration took a much firmer stand against such a development and nuclear proliferation in general.[81] Although the United States faced substantial difficulties maintaining allied unity on the question—not only West Germany but also France objected—the U.S. position in favor of preventing West Germany from acquiring an independent nuclear weapons capability remained consistent in negotiations with the Soviet Union from 1961 through 1963.[82]

Was the U.S. Deterrent Message Credible and Convincing?

While U.S. deterrent messages in this second Berlin crisis were relatively clear, Khrushchev did not initially consider these promises to fight to stay in Berlin fully credible. The credibility of U.S. threats was gradually established over the course of the crisis, contributing to its eventual peaceful resolution. Before surveying the chronological evolution of U.S. credibility, we will first review the structural factors that shaped both Soviet and U.S. perceptions and options in the crisis.

The credibility of U.S. promises to remain in West Berlin during the crisis of 1958–1962, and to go to war if the Soviets challenged their position, ultimately rested on Soviet perceptions of U.S. willingness to use nuclear weapons in any resulting conflict. In the event of a conflict, potentially hundreds of thousands of nearby East German and Soviet troops would have relatively quickly overrun the roughly 11,000 Western troops in Berlin, and neither side anticipated a meaningful Western conventional counterattack.[83] Instead, U.S. contingency plans established under Eisenhower for a Berlin crisis in which access to the city was interrupted called for a probe in force to attempt to reach the city with a single division, but if that force were attacked, to then resort to general war, including the use of nuclear weapons.[84] The available conventional response options left the incoming Kennedy administration dissatisfied.[85] However, the flexible response they preferred, involving increases

---

[80] Kempe, 2011, p. 447.

[81] Trachtenberg, 1999, pp. 281, 284.

[82] Trachtenberg, 1999, pp. 329, 332, 344–345, 386–390.

[83] Kempe, 2011, pp. 55–56, 385.

[84] Kempe, 2011, pp. 442–433.

[85] Kempe, 2011, p. 53.

in overall U.S. and NATO conventional capabilities, did not exist during the 1961 crisis. U.S. leaders, including Kennedy, feared they would be confronted with a remarkably short road between initial, limited hostilities and a decision to escalate to general nuclear war.[86]

Two factors enhanced the credibility of U.S. willingness to resort to nuclear weapons to defend a small, isolated outpost such as Berlin. First, the experience of the 1948 crisis, in which the United States signaled that it was able and willing to deploy nuclear weapons in order to defend its position in the city, even though not ultimately tested, established a baseline expectation regarding U.S. policy and redlines.[87] Second, in this period the United States retained a substantial advantage over the Soviet Union in the number and deliverability of nuclear weapons.[88] Although each side had the ability to strike the other's territory with nuclear weapons, the American ability to do so using much larger numbers of ICBMs, strategic bombers, and sea-launched missiles was apparent.[89] This disparity raised the prospect that the United States could still seek to win a nuclear exchange with the Soviet Union in this period, although certainly with much greater risk to the U.S. homeland than in 1948.

Despite these advantages, Soviet perceptions of U.S. resolve varied across two different issues, and over time. Khrushchev was initially willing to threaten Kennedy with war over Berlin because he assessed that the new president was likely to back down rather than initiate hostilities.[90] Khrushchev's threats were therefore, from his perspective, not overly risky.[91] As the crisis developed, however, Kennedy was able to signal credibly that he would be willing to fight over Berlin.[92] Kennedy's speech on July 25, 1961, clearly stated that the United States would defend Berlin and linked the city's security with NATO and the United States, and also included an announcement that the United States would substantially increase its long-term investment in conventional forces.[93] After the construction of the Berlin Wall in August, the United States also

---

[86] As will be discussed, even once flexible response became U.S. policy, the number of U.S. forces in Germany did not increase substantially, although their capabilities did become more robust. See Kugler, 1991, p. 104.

[87] Miller, 1998, pp. 24–25.

[88] Kugler, 1991, p. 9.

[89] Kugler, 1991, p. 9. Indeed, the Soviets viewed their own ICBM capabilities at the time as existing only "on paper." See Taubman, 2003, p. 504.

[90] This perception was enhanced by Khrushchev's assessment of Kennedy's refusal to engage directly in the Bay of Pigs in spring 1961. See Petr Lunák, "Khrushchev and the Berlin Crisis: Soviet Brinkmanship Seen from Inside," *Cold War History*, Vol. 3, No. 2, January 2003, p. 54; Taubman, 2003, p. 495; and Kempe, 2011, p. 307.

[91] Kempe, 2011, p. 25.

[92] Lunák, 2003, p. 78; Smyser, 2009, p. 92. In part, Khrushchev's escalating concern that Kennedy really would go to war also reflected a concern that the young President may not have been able to stand up to hawkish elements in the United States. See Taubman, 2003, p. 502.

[93] Lunák, 2003, p. 74, Donald A. Carter, *The U.S. Military Response to the 1960–1962 Berlin Crisis*, Washington, D.C.: U.S. Army Center of Military History, 2011, p. 2; Kempe, 2011, p. 314.

sent a reinforced battle group via a land access route to West Berlin.[94] The United States further underlined its resolve through decisions in October 1961 to send additional forces to Europe, including several U.S. Air Force squadrons and pre-positioned equipment for two ground divisions, as well as by public speeches by U.S. defense officials emphasizing the survivability of the U.S. nuclear deterrent and the intent to employ it if necessary to safeguard the U.S. position in Berlin.[95]

Khrushchev's approach to the crisis remained bellicose throughout—including, of course, the building of the wall, as well as military exercises that simulated the use of Soviet tactical nuclear weapons in a conflict over Berlin.[96] However, by the middle of August he did back away from threatening continued Western presence in and access to Berlin directly.[97] Further, while Soviet military officials prepared for the possibility of a limited U.S. conventional effort to fight through a potential Soviet blockade of Berlin, they were not ready for any further escalation to general war.[98] Despite backing away from direct threats to the Western position, the Soviet perception that the United States was considering tearing down the wall may have complicated the Soviet decision to maintain a publicly confrontational stance. The Soviets felt compelled to signal that such an escalatory action would lead to conflict.

Yet, faced with an increasingly credible and resolute U.S. position by October 1961, Khrushchev ultimately backed down and proved unwilling to risk even a low-level conflict over the city once the wall had been built and accepted.[99] While he may have had initial doubts over U.S. willingness to fight to stay in Berlin, once he became convinced that a fight was possible he could not be certain that the United States would not be willing to use nuclear weapons.[100] Indeed, the lack of viable conventional options with which the United States could respond to a conflict in Berlin made the U.S. choice one between humiliating defeat and nuclear escalation. That nuclear risk proved sufficiently credible to Khrushchev to discourage further escalation of the crisis in Berlin.[101] It is worth nothing, though, that Soviet questions about

---

[94] Gelb, 1986, p. 227; Carter, 2011, p. 6.

[95] Kempe, 2011, pp. 438, 446.

[96] Kempe, 2011, p. 382.

[97] Kempe, 2011, pp. 244–246, 424.

[98] Taubman, 2003, p. 504.

[99] Lunák, 2003, p. 77; Taubman, 2003, pp. 537–538; Smyser, 2009, p. 92; Kempe, 2011, pp. 412, 429, 481.

[100] Smyser, 2009, p. 141. Some Americans did, however, question U.S. willingness to do so, including Henry Kissinger, who was acting as an outside consultant to the Kennedy administration in summer 1961. Kennedy, however, appears to have spent much of 1961 preparing himself to use nuclear weapons over Berlin if a conflict broke out, demanding detailed assessment of the procedures and likely consequences of doing so. See Kempe, 2011, pp. 304, 434.

[101] Notably, this same logic was used by figures such as NATO SACEUR Lauris Norstad, West German chancellor Konrad Adenauer, and French president Charles de Gaulle to argue against the Kennedy administration's plans to implement a flexible response, providing more conventional options to respond to Soviet aggression in

the credibility of U.S. resolve to use force appear to have remained sufficient to allow Khrushchev to test Kennedy again in Cuba the following year.

1964–1989

How Motivated Was the Soviet Union to Eject the Western Allies from Berlin?

Throughout the later Cold War period, the Soviet Union appears to have had limited motivation to force the United States and its allies out of Berlin. After the building of the wall, the city ceased to serve as a substantial transit corridor for emigration to the West. While the United States did increase the capabilities of its conventional forces in Europe during this period, West Germany remained a nonnuclear state, with the nuclear weapons that were in the country remaining under American control. Furthermore, at least through the mid-1970s, a general détente prevailed between the East and the West in Europe, as the 1971 Four Power Agreement on Berlin and the 1975 Helsinki Accords demonstrated.[102] The key strategic issues that had motivated the Soviets to target Berlin in earlier crises did not become acute in this period. Moreover, even if the Soviet Union had succeeded in taking Berlin in a limited conflict, this would have represented the end of the credibility of U.S. extended deterrence in Europe for NATO allies, a development that would not have been entirely advantageous to the Soviet Union. The United States would then have lost any leverage it had to keep West Germany from developing an independent nuclear deterrent, a key Soviet fear.[103]

Was the U.S. Deterrent Message Clear and Explicit?

The U.S. commitment to the defense of Berlin remained consistent in this period. The U.S. troop presence in the city, the Berlin Brigade, continued to provide clear and public evidence of U.S. commitment to the defense of West Berlin.[104] Following Kennedy's famous visit to the city in 1963, U.S. presidents of both parties repeatedly visited West Berlin, including Richard Nixon in 1969, Jimmy Carter in 1978, and Ronald Reagan in 1982 and 1987, tying the prestige of the president with the continued defense of the city.[105]

Germany. They argued that introducing greater conventional forces would only serve to raise questions in the minds of the Soviets regarding whether the United States would ultimately be willing to employ nuclear weapons in Europe or whether it would instead try to keep the conflict purely conventional, an area where the Soviet Union would regardless maintain a clear advantage. See Kempe, 2011, p. 441.

[102] Doeker, Melsheimer, and Schroder, 1973, p. 54; Judt, 2005, p. 501.

[103] Trachtenberg, 1999, p. 399.

[104] Robert P. Grathwol and Donita M. Moorhus, *American Forces in Berlin: Cold War Outpost, 1945–1994*, Washington, D.C.: U.S. Department of Defense, Legacy Resource Management Program, 1995, p. 120.

[105] U.S. Department of State, U.S. Embassies and Consulates in Germany, "Visits of U.S. Presidents to Germany Since 1945," n.d.

Was the U.S. Deterrent Message Credible and Convincing?

The credibility of the U.S. deterrent message in Berlin in the late Cold War period rested on the same factors that helped to end the earlier crises. While the U.S. troops in Berlin did modestly increase their capabilities after the 1961 crisis, even the members of the Berlin Brigade acknowledged they anticipated being quickly overrun in the event of a conflict.[106] Instead, U.S. deterrence in this period relied on the U.S. promise, credibly signaled in the earlier crises, that a conflict over Berlin would result in general war with the United States and that in the event of such a war the Soviet Union could not assume that escalation could be controlled and strategic nuclear weapons not employed. Given the tremendous strategic costs to the United States from simply abandoning its position in Berlin and the threat that would pose to its position in Europe more broadly, the Soviets viewed U.S. threats to escalate to general war as sufficiently credible not to risk further crises for limited gains.

Implications for the Framework

In the end, U.S. efforts to deter the Soviet Union from attacking West Berlin during the Cold War succeeded. Two acute crises did erupt over the city in 1948 and 1961 in which the Soviet Union tested the credibility of U.S. commitments. These crises arose in large part due to Soviet insecurity over developments in Germany more broadly, but the choice of Berlin as a potential pressure point for the West also reflected Soviet perceptions that U.S. resolve to fight to stay in Berlin was at least initially uncertain. By the later Cold War period, after the demonstration of Western resolve in the two crises and the establishment of a modus vivendi between the two sides over Germany, the potential high costs and low benefits of further crises became clear and the Western position in the city persisted largely unchallenged.

The case of West Berlin has implications for each of the three main categories of variables affecting the success of deterrence in the framework developed in this report. From 1948 to 1963, the Soviet Union had relatively clear and strong motivations for ejecting the United States and its allies from Berlin, or at a minimum threatening to do so in pursuit of additional strategic imperatives. The Western enclave inside East Germany functioned as a threat to the Soviet satellite, providing a route for mass emigration and economic influence. It was also a convenient point by which to exert pressure on the United States to achieve Soviet goals in West Germany that threatened Soviet security more broadly—first to arrest the formation of the West German state, and then to keep that state from developing nuclear weapons. Despite these motivations, however, Soviet leaders never intended to use even limited levels of force over Berlin. While they were willing to accept heightened levels of risk of inadvertent conflict

---

[106] Grathwol and Moorhus, 1995, pp. 120, 122.

in order to achieve their objectives, at no point did they intend to try to take the city by force and precipitate a war with the United States. While they may have ideally preferred to kick the West out of Berlin, proximate Soviet perceptions of U.S. commitments and capabilities shaped their goals. Soviet motivations to threaten the Western position in Berlin, although clearly affected by the Soviet Union's perceptions of U.S. deterrence, should still be distinguished from other cases where adversaries had a clear intent and plan to initiate hostilities. Berlin therefore does not test the ability to deter adversaries bent on attack, but it does demonstrate how deterrence can be used to prevent adversaries from considering such a course of action to begin with, despite clear incentives.

The clarity and credibility of U.S. threats to fight to stay in Berlin varied. In 1948, despite facing a United States that retained a nuclear monopoly, the Soviet Union assessed that it could simply cut access to the city and force the United States to withdraw and to accede to Soviet demands, or itself be the one to initiate hostilities, a prospect which the Soviets judged to be unlikely given the massive Soviet conventional superiority in Germany. At the time, senior U.S. officials seriously considered abandoning Berlin (although President Truman was not among them). While the United States took steps to bolster the credibility of its willingness to go war if attacked in Berlin in 1948, as with the deployment of B-29 bombers to Europe, the Soviets never planned to force the crisis to the point of armed conflict. The unexpected success of the Western airlift in providing an alternative way out of the dilemma the Soviets had constructed helped to buy the United States a reprieve.

By 1961, despite substantial NATO investments, Soviet forces continued to dwarf NATO conventional forces. Meanwhile, the Soviets had developed their own nuclear deterrent, which while still smaller and less easily deliverable than U.S. nuclear weapons had become a substantial threat, including to the U.S. homeland. Against this backdrop Khrushchev assessed that whatever nuclear edge the United States might retain, Kennedy was willing to employ it over a crisis in Berlin. While U.S. public statements regarding their determination to fight to retain their position in Berlin had remained clear, they were not initially credible to the Soviets. In part, this appears to have stemmed from Kennedy's refusal to intervene in the Bay of Pigs fiasco earlier that year, which Khrushchev assessed as a sign of weakness. However, the lack of initial U.S. credibility also reflected the inherent difficulty of signaling U.S. willingness to risk nuclear war in order to maintain its position in an exposed enclave that could not long defend conventionally. Establishing this credibility required clear demonstrations of U.S. intent, including deployment of additional forces to Berlin and Western Europe during the crisis, as well as repeated, explicit public statements by Kennedy and other senior administration officials that conflict in Berlin meant general war with the United States; these statements would have extracted a devastating diplomatic and political cost to the speakers. To be effective, U.S. threats did not need to guarantee a nuclear response to a conventional attack in the minds of the Soviet leadership. To outweigh the relatively limited Soviet goals in Berlin, particularly after the wall was built, they had only to make such a response plausible enough given the existential risks to the Soviets that would be involved.

The less eventful success of deterrence in the later Cold War period can be tied to two factors. First, the enhanced credibility of U.S. deterrent promises was buttressed by the commitment demonstrated in 1961, as well as the additional investment of resources in the defense of Western Europe more broadly. Second, Soviet motivation to threaten the city lessened once the wall was constructed and the United States had tacitly promised to link the territorial status quo with pressure on the West Germans to remain a nonnuclear state. Both factors were crucial in ensuring that general deterrence continued to hold for the remainder of the Cold War.

The successful resolution of the 1948 and 1961 crises showed the necessity of clear, credible U.S. commitments for deterrence to work, but it also highlights the importance of tactical flexibility and recognition of the legitimate security concerns of the other party. In the 1961 crisis, in particular, the United States was able to preserve its position in Berlin and in Europe more broadly, but the Soviets did not come away empty-handed. U.S. acquiescence to the building of the Berlin Wall, and more generally to granting the Soviets a free hand to run East Berlin as they saw fit, were seen at the time as signs of weakness and lack of resolve that could embolden future Soviet aggression, for they represented a rollback of U.S. rights under postwar agreements. However, Kennedy judged that Khrushchev's need to address migration flows through the city was acute, and Khrushchev's motivation to act to alter the status quo strong. Kennedy was therefore willing to accept a weakening of U.S. rights in the city in order to allow the Soviets to stabilize the situation in a manner consistent with continued Allied presence in West Berlin. Executing this limited retreat from previous U.S. positions while enhancing the credibility of U.S. promises to retreat no further required numerous signaling efforts, including explicit public commitments and military movements. In the end, though, this strategy (combined with the related understanding that the United States would prevent an independent West German nuclear capability) proved effective in limiting Soviet motivations to again risk war over Berlin. At the same time, it maintained the clarity and credibility of U.S. commitments to fight over the city.

This case therefore highlights the types of steps that may be required for the United States to make clear and credible commitments to extended deterrence. While local conventional military superiority was not required in this case, a credible willingness to escalate to general war was. Establishing this willingness required costly U.S. signals regarding political and military commitments. At the same time, a blind refusal to consider any modifications to previous commitments would likely have increased the risk deterrence failure and conflict in 1961 given the security concerns the Soviet Union faced. Efforts to ensure the clarity and credibility of U.S. commitments also need to be considered in light of the effect that they may have had on adversary motivations.

APPENDIX C

Qualitative Case Study Analyses: Deterring Saddam, 1990

Iraq's invasion of Kuwait on August 2, 1990, is commonly viewed as a textbook failure of deterrence by the administration of President George H. W. Bush. The invasion came after several months of Iraqi leader Saddam Hussein's repeated threats against his Persian Gulf neighbors. Beginning in May 1990, Saddam called for Kuwait and other Arab states in the gulf region to cut back on what he deemed to be their overproduction of oil, which had brought down the price and placed Baghdad under enormous financial pressure. At the time, Iraq required a price of $18 per barrel to be able to pay off $80 billion in debts it had incurred during its eight-year war with Iran.[1] Saddam insisted that his fellow Arabs should appreciate his service in protecting them from Ayatollah Khomeini, and that they should show their gratitude by helping raise oil prices.[2] Saddam was also upset by what he alleged to be Kuwait's theft of Iraqi oil through horizontal drilling along the border. By summer 1990, Iraq's rhetoric had become even more heated. On July 16, Iraqi foreign minister Tariq Aziz sent a letter to the Arab League claiming that Kuwait's refusal to cancel Iraq's war debts, its overproduction of oil, and its lack of interest in resolving the border dispute were all tantamount to military aggression.[3] The next day, Saddam made a speech alleging that the overproduction of oil was part of an anti-Iraq conspiracy perpetrated by the United States and its Arab allies. Several days later, Saddam deployed the Republican Guard to Basra. By the eve of the August 2 invasion, eight Republican Guard divisions (120,000 fighters and 1,000 tanks) had been stationed along the Kuwaiti border.[4] Despite all these warning signs, most U.S. troops and warplanes were over 7,000 miles away.

---

[1] Baghdad owed around $40 billion to fellow Arab countries and another $40 billion to the West. See Stein, 1993.

[2] "Oral History: Tariq Aziz," 1995.

[3] Efraim Karsh and Inari Rautsi, *Saddam Hussein: A Political Biography*, New York: The Free Press, 1991.

[4] Gordon and Trainor, 1995, pp. 3–54.

The Bush Administration's Decisionmaking

Various studies have offered multiple explanations as to why the Bush administration failed to deter Saddam from occupying Kuwait. Some argue that Bush was too wedded to his goal of improving relations with Saddam. Others claim that U.S. officials projected their own rationalizations onto Saddam and failed to understand his objectives and threat perception. In addition, Washington's desire to accommodate the wishes of other regional leaders is thought to have played a role in leaving the United States unprepared. Finally, in the lead-up to the invasion, the Bush administration appears to have been distracted by numerous other global issues.

President Bush inherited his approach to the Baghdad regime from the administration of President Ronald Reagan, which viewed Saddam as the bulwark preventing the region from falling under the influence of the newly established Islamic Republic of Iran. As such, the United States restored relations with Iraq in 1984 (Iraq had cut ties in 1967 following U.S. support for Israel during the Six-Day War), and tilted toward Saddam during most of the Iran-Iraq War.[5] In October 1989 the Bush administration issued NSD 26, which would serve as the guideline for U.S. policy in the Persian Gulf until Saddam's invasion of Kuwait.[6] The directive claimed that normalized relations with Iraq were in the U.S. national interest. To maintain relations with Iraq, the United States would need to moderate Saddam's behavior by providing Iraq with economic incentives and pursuing opportunities for U.S. firms to help in Iraq's postwar reconstruction. According to then–Deputy National Security Adviser Robert Gates, U.S. officials did not expect Saddam would change dramatically; they did hope, however, that he could become a more predictable dictator like Syria's Hafez al-Assad.[7] Meanwhile, NSD 26 also declared that U.S. access to Persian Gulf oil and preserving the security of regional friendly states were vital to national security, and that Saddam should be made to understand that any further use of chemical or biological weapons would result in sanctions. (For a timeline of events leading up to the invasion, see Figure C.1.)

Projecting their frame of mind onto Saddam, most U.S. officials failed to comprehend fully the threat Iraq posed to Kuwait. They assumed that Saddam would refrain from aggressive behavior, and focus instead on reconstructing his country in the wake of a costly war with Iran.[8] But while Iraq had cut its forces by half following its cease-fire with Iran, Saddam's 400,000-man army was still the largest in the region.[9] Furthermore, the widespread belief that "Arab countries did not invade other

---

[5] Zachary Karabell, "Backfire: US Policy Toward Iraq, 1988–2 August 1990," *Middle East Journal*, Vol. 49, No. 1, 1995, pp. 28–47.

[6] National Security Directive 26, October 2, 1989.

[7] Karabell, 1995.

[8] Richard N. Haass, *War of Necessity, War of Choice: A Memoir of Two Iraq Wars*. New York: Simon and Schuster, 2009, 60ff.

[9] "Oral History: Tariq Aziz," 1995.

Figure C.1
Timeline of Events in 1990 Leading Up to the Iraqi Invasion of Kuwait

- February 15: A Voice of America article calls for overthrow of Arab dictatorships. Saddam considers this to be a statement of official U.S. policy.
- April: The United States suspends $500 million in agriculture credits in response to Saddam's threats against Israel.
- May: Saddam addresses the Arab Summit, calling for other regional countries to stop overproduction of oil.
- Late June: Saddam first discusses the invasion of Kuwait with subordinates (according to Tariq Aziz).
- July 10: Oil ministers of Iraq, Kuwait, Saudi Arabia, Qatar, and the UAE agree to limit production to increase price of oil; however, several days later, Kuwait announces that it would only abide by the quota until the fall.
- July 16: Foreign Minister Tariq Aziz sends letter to Arab League claiming that Kuwait's actions equate to military aggression: refusing to cancel Iraq's debts, surpassing OPEC production quota, and not resolving the border dispute.
- July 17: Saddam makes a speech accusing the United States, Zionists, and their Arab allies of conspiring against Iraq by increasing oil production.
- July 18: The U.S. State Department states that the United States is committed to defending allies in the Persian Gulf and ensuring the free flow of oil through the Strait of Hormuz.
- July 21: The Republican Guard is mobilized in Basra.
- July 23: The United States sends two KC-135 aerial tankers and a C-141 cargo transport plane to the UAE and moves six warships closer to Kuwait. Meanwhile, the White House denies that U.S. vessels in the Persian Gulf have been put on alert.
- July 24: State Department Spokesperson Margaret Tutwiler says the United States and Kuwait do not have a defense treaty.
- July 25: Ambassador April Glaspie is summoned to meet Saddam. Glaspie tells Saddam that Washington has "no opinion on the Arab-Arab conflicts, like your border disagreement with Kuwait."
- July 25: CIA analyst Charles Allen warns the White House of 60% likelihood that Iraq will invade Kuwait. His warning is dismissed by senior officials.
- July 28: Bush sends Saddam a note that "we still have fundamental concerns about Iraqi policies and activities, and we will continue to raise these concerns with you in a spirit of friendship and candor."
- July 28: Despite White House opposition, the Senate and House pass sanctions on Iraq for human rights violations and its aggressive behavior in the region.
- July 31: A National Intelligence Council memorandum judges Iraqi attack on Kuwait likely, but full occupation unlikely. Iraq-Kuwait talks begin in Saudi Arabia.
- August 1: Eight Republican Guard divisions amass along the Kuwaiti border. Most U.S. troops and warplanes are over 7,000 miles away. U.S. Assistant Secretary of State John Kelly calls in Iraqi ambassador to express that things have become "extremely serious."
- August 2: Iraq invades Kuwait.

Arab countries" blinded Washington to the possibility of an Iraqi invasion.[10] Therefore, all of Iraq's aggressive actions and rhetoric were interpreted merely as bluffs to gain concessions from Kuwait. A July 21, 1990, alert from the Defense Intelligence Agency (DIA) reflected this belief, assessing that Iraqi troop mobilization along the Kuwaiti border was "probably a continuation of Saddam Hossein's [sic] campaign of

---

[10] Gordon and Trainor, 1995, p. 5.

intimidation and force posturing designed to raise oil prices and end cheating on oil production quotas."[11]

On July 25, when Saddam summoned U.S. Ambassador April Glaspie to his palace and railed against Kuwait and the United Arab Emirates (UAE) for engaging in "economic warfare," Glaspie reported back to Washington that the United States should refrain from publicly criticizing Saddam's actions because "He does not want to further antagonize us."[12] According to Glaspie, recent U.S.-UAE military exercises had been sufficient in sending a message to Saddam that aggression would not stand. Meanwhile, members of the Bush administration and officials from the Organization of the Petroleum Exporting Countries (OPEC) assured reporters that Saddam was simply trying to scare Kuwait and the UAE into cutting back on their oil production.[13] By July 31, however, the National Intelligence Council assessed that an Iraqi attack on Kuwait was likely. Again, however, the belief was that the attack would be limited to the seizure of the Rumaila oil field and some islands—all in order to gain leverage in Iraq-Kuwait talks. Led by these assumptions, the United States did not strongly encourage the Kuwaitis to make concessions to Saddam during the Iraq-Kuwait talks in Jeddah that began on July 31.[14]

Arab leaders also appear to have influenced the Bush administration's strategy regarding the Iraq-Kuwait dispute. For most of 1990 neither the Egyptians, Kuwaitis, nor Saudis believed that Saddam would invade Kuwait.[15] By the summer, American and Kuwaiti officials thought that the worst-case scenario would be that Iraq would occupy the border area and the Bubiyan and Warba Islands for a limited time in order to use them as bargaining chips.[16] The Kuwaitis rebuffed a U.S. offer to conduct joint military exercises out of concern that such exercises would needlessly antagonize Saddam.[17] Moreover, most Arab allies, including Jordan, requested that Washington stay out of what they deemed an inter-Arab dispute.[18] Senior administration officials therefore believed that the Arab street (and the American public) would oppose any U.S. decision to send in troops to deter Saddam or to repel him if he were to take only

---

[11] Gordon and Trainor, 1995, p. 17.

[12] Gordon and Trainor, 1995, p. 22.

[13] Youssef Ibrahim, "Iraq Said to Prevail in Oil Dispute with Kuwait and Arab Emirates," *New York Times*, July 26, 1990.

[14] Stein, 1993.

[15] James A. Baker III, *The Politics of Diplomacy: Revolution, War and Peace, 1989–1992*, New York: G. P. Putnam's Sons, 1995.

[16] F. Gregory Gause III, "Iraq and the Gulf War: Decision-Making in Baghdad," unpublished manuscript, n.d.

[17] Haass, 2009.

[18] One exception was the UAE defense chief, Sheikh Mohammad Bin Zayed, who asked the United States to send two refueling tankers to extend the capabilities of UAE Mirage fighters. See Gordon and Trainor, 1995.

small parts of Kuwait.[19] According to Janice Gross Stein, "These political judgments made an effective strategy of deterrence virtually impossible."[20]

Finally, the Bush administration appears to have been distracted by potentially larger crises prior to Iraq's invasion. For instance, Robert Gates and other White House officials had recently visited New Delhi in an effort to defuse India-Pakistan tensions that they feared could lead to nuclear war.[21] And according to a senior Bush administration official, the Iraq crisis "came in a bit of a vacuum, at a time when everyone was focusing on German reunification."[22] Additionally, according to Zachary Karabell, Bush's Iraq policy "backfired because fear of the long-term fundamentalist threat posed by Iran largely obscured the short-term threat posed by Saddam Hussein."[23]

But the warning signs had been there all along. Following the Iraqi occupation of Kuwait, Central Intelligence Agency (CIA) analysts looked back on Saddam's past actions, including his invasion of Iran, and assessed that he was impulsive and that he often miscalculated. A 1979 Pentagon report had stated that the United States would need to exhibit a strong willingness to use force in order to protect Iraq's neighbors, as Baghdad had a history of claiming Kuwait as Iraqi territory, and Kuwaiti oil fields along the coastline were an attractive target for Saddam.[24] Some analysts had also warned of a potential attack. Most prominent among them was the CIA's national intelligence officer for warning, Charles Allen. On July 25, Allen issued a warning to the White House assessing that the mobilization of Iraqi troops indicated a high likelihood that Iraq would invade Kuwait.[25] Senior officials dismissed his warning, along with those issued by DIA senior Iraq analyst Walter Lang. The cognitive dissonance had been so strong as to blind most officials to the true nature of the threat; taking warnings of an invasion seriously would have meant that the Reagan and Bush administrations' policies had been gravely misguided.

The Lack of Clear Signals to Saddam

Because of the Bush administration's faulty assumptions, the United States did not issue clear and strong warnings to Saddam regarding the costs he would incur should he invade Kuwait. According to Stein, "The diplomacy of deterrence in the critical two

---

[19] Stein, 1993.

[20] Stein, 1993, p. 131.

[21] Haass, 2009.

[22] Elaine Sciolino and Michael R. Gordon, "U.S. Gave Iraq Little Reason Not to Mount Kuwait Assault," *New York Times*, September 23, 1990.

[23] Karabell, 1995, 47.

[24] Gordon and Trainor, 1995.

[25] "The Gates Hearings; Early Indicators of Kuwait Invasion," *New York Times*, September 25, 1991.

weeks preceding the invasion was inconsistent, incoherent, and unfocused."[26] Bush's mixed messages to Saddam are the most commonly cited factor in explaining Washington's failure to deter the Iraqis. Several months after the Iraqi invasion, even Secretary of State James Baker contended that the occupation could "absolutely" have been prevented if the United States had issued strong warnings to Saddam.[27]

Prior to the invasion, however, officials appear to have felt that they were issuing an adequate balance of sticks and carrots to keep Saddam in line. After he threatened in April 1990 that Iraq would "make fire eat up half of Israel," the United States suspended $500 million in agriculture credits.[28] In late May, senior National Security Council staffer Richard Haass was dispatched to Baghdad to warn Tariq Aziz that Iraq's actions were causing concern in Washington and could impact whether the United States saw Iraq as an adversary. On July 18, the State Department reiterated U.S. determination to ensure the free flow of oil through the Strait of Hormuz and to "remain strongly committed to supporting the individual and collective self-defense of our friends in the Gulf with whom we have deep and longstanding ties."[29] And on July 23, 1990—as Saddam's rhetoric was heating up, and following a plea from the government of Abu Dhabi—the United States sent two KC-135 aerial tankers and a C-141 cargo transport plane to the UAE. It also moved six warships closer to Kuwait. Bush administration officials apparently felt that this action, along with joint exercises with the UAE, would make it clear to Saddam that the United States would support its Gulf Arab allies, and that it was willing to use force to ensure the free flow of oil through the Strait of Hormuz.[30]

However, critics later claimed that the Bush administration undercut these warnings by sending unnecessarily appeasing messages to Saddam. For instance, when in April 1990, Bush sent five senators to meet with Saddam in Baghdad to discuss concerns over his pursuit of chemical and nuclear weapons, they also took pains to assure him that the harsh criticism he was receiving in the American press did not directly reflect the White House's position.[31] Later, following the shipment of military equipment to the UAE, the White House publicly denied that U.S. vessels in the Persian Gulf had been put on alert.[32] In a July 24 press conference, State Department spokesperson Margaret Tutwiler reiterated that the United States and Kuwait did not

---

[26] Stein, 1993, p. 126.

[27] Leslie H. Gelb, "Mr. Bush's Fateful Blunder," *New York Times*, July 17, 1991.

[28] Karabell, 1995, p. 39.

[29] Haass, 2009, p. 56.

[30] Michael R. Gordon, "U.S. Deploys Air and Sea Forces After Iraq Threatens 2 Neighbors," *New York Times*, July 25, 1990a.

[31] Karsh and Rautsi, 1991, pp. 194–217.

[32] This strategy was likely influenced by Chairman of the Joint Chiefs of Staff Colin Powell, who opposed using the military to send diplomatic messages. See Gordon, 1990a; and Gordon and Trainor, 1995.

have a formal defense treaty.³³ It is thus unsurprising that, shortly before the invasion, Kuwaiti officials expressed concern that the United States had not made an explicit, public promise to protect the Kuwaitis from an attack.³⁴ A personal note Bush sent Saddam on July 28 stated that "we still have fundamental concerns about certain Iraqi policies and activities, and we will continue to raise these concerns with you *in a spirit of friendship and candor*."³⁵ But most damning for the administration was Ambassador Glaspie's assurance in her July 25 meeting with Saddam that the United States had "no opinion on the Arab-Arab conflicts, like your border disagreement with Kuwait."³⁶ Yet according to Robert Gates, by that point, "Barring an ironclad threat to oppose Hussein by force, there was little Glaspie could have done or said that would have made a difference."³⁷ And although on August 1, Assistant Secretary of State John Kelly did send a forceful message to the Iraqi ambassador in Washington that things had become "extremely serious," Saddam had already decided on a full occupation of Kuwait.³⁸

Saddam's Threat Perception

While throughout the 1980s U.S. officials thought that they were improving relations with Iraq, Saddam had increasingly become convinced that the United States was bent on undermining him. Therefore Saddam saw his invasion of Kuwait as a necessary action in the face of a conspiracy to overthrow him on the part of Israel, Kuwait, Saudi Arabia, the UAE, and the United States.

According to tape recordings of Saddam and his advisers (captured by U.S. forces following the 2003 invasion of Iraq), as well as testimony from his foreign minister, Tariq Aziz, the 1986 Iran-Contra Affair served as a watershed moment for Saddam.³⁹ The incident convinced him that the United States opposed him. In a December 1990 meeting with his advisers, Saddam said, "The war was launched on us long before all of this. It officially started in the 1986 meeting, and was exposed under the

---

[33] Karabell, 1995.

[34] Michael R. Gordon, "Iraq Army Invades Capital of Kuwait in Fierce Fighting," *New York Times*, August 2, 1990b.

[35] Michael R. Gordon, "Pentagon Objected to Bush's Message to Iraq," *New York Times*, October 25, 1992, emphasis added.

[36] Barry R. Schneider, *Deterrence and Saddam Hussein: Lessons from the 1990–1991 Gulf War*, Maxwell Air Force Base, Ala.: U.S. Air Force Counterproliferation Center, 2009, p. 13.

[37] Karabell, 1995, p. 45.

[38] Stein, 1993, p. 126.

[39] Central Intelligence Agency, *Comprehensive Report of the Special Advisor to the DCI on Iraq's WMD*: Vol. 1, Washington, D.C.: Central Intelligence Agency, 2004.

title 'Irangate.' . . . August the second [the occupation of Kuwait] was an attack and a defense both at the same time."[40]

In Saddam's mind, the reason for U.S. opposition to him was that Washington did not want a strong independent Arab nation that could prevent the Americans from dominating the Persian Gulf. Saddam believed that because the Soviet Union had fallen, the United States now believed it should have a free hand to dominate the Middle East.[41] He became convinced that the CIA was trying to overthrow him. There reportedly had been failed coup attempts against Saddam in September 1989 and January 1990.[42] Therefore, by 1990, Saddam may have interpreted all efforts by the Bush administration to improve relations to be empty rhetoric, and therefore analyzed all U.S. actions through his paranoid prism. When a February 1990 Voice of America opinion piece called for the overthrow of Arab dictatorships, Saddam interpreted it as a statement of official White House policy—despite Ambassador Glaspie's reassurance that the administration had no control over what the media published. In April 1990, when the U.S. Congress suspended agriculture credits to Iraq, the Iraqi leadership saw this as a White House attempt to destabilize the government.[43] This perception endured despite the Bush administration's opposition to hard-liners in Congress. Saddam also saw the U.S. naval presence in the Persian Gulf as a direct threat. Reflecting this belief, Aziz asked Haass what other reason there would be for the ships to remain since the Iran-Iraq War was over and there was no longer a need to protect oil tankers.[44] According to an Iraqi transcript of his July 25 meeting with Glaspie, Saddam accused the United States of supporting "Kuwait's economic war against Iraq."[45]

Finally, what made the dispute with Kuwait so critical was that without adequate oil revenue to reconstruct Iraq, Saddam feared domestic unrest.[46] This fed into his larger concern that the fall of the Soviet Union would inspire Iraqis to try to overthrow him.[47]

Saddam's Decision to Occupy Kuwait

According to interviews conducted with those from his inner circle, Saddam appeared to have been considering the Kuwait invasion for several months prior to discussing it

---

[40] Kevin M. Woods, David D. Pakki, and Mark E. Stout, eds., *The Saddam Tapes: The Inner Workings of a Tyrant's Regime, 1978–2001*, Cambridge: Cambridge University Press, 2011, p. 35.

[41] "Oral History: Tariq Aziz," 1995.

[42] Gause, n.d.

[43] Stein, 1993.

[44] Haass, 2009.

[45] Karsh and Rautsi, 1991, p. 215.

[46] "Oral History: Tariq Aziz," 1995.

[47] Gause, n.d.

with his subordinates.⁴⁸ According to Aziz, the end of June 1990 was the first time that Saddam brought up his idea to invade.⁴⁹ On July 10, Kuwait and the other Gulf States agreed to limit oil production to increase the price of oil to the $18 per barrel that Iraq needed. But several days later, Kuwait announced that it would hold to the agreement for only a few months. Congress's July 28 passage of new sanctions on Iraq for human rights violations and its aggressive behavior in the region likely heightened Saddam's perception of an impending threat.⁵⁰ According to a prominent media executive, Saad al-Bazzaz, Saddam decided the next day that Iraq would fully occupy, rather than simply invade, Kuwait.⁵¹ According to Aziz, Saddam's decision to occupy was based on the assumption that the United States would retaliate regardless of whether he simply invaded or fully occupied.⁵² Therefore, Iraq might as well fully occupy Kuwait in order to prevent U.S. ground forces from amassing there. In the two weeks prior to the invasion, the Iraqi media lessened their attacks on Kuwait, apparently in an attempt to give its rivals a false sense of security.⁵³ The final decision on the timing of the occupation was made on August 1, after the Iraqi delegation returned empty-handed from talks with Kuwait in Saudi Arabia and what they deemed to be Kuwait's "arrogance" during the negotiations.⁵⁴

It is also clear that Saddam's poor reasoning process played a large role in his decision to occupy Kuwait. First, lack of dissent among his advisers resulted in faulty conclusions. Saddam had "created and ruthlessly enforced a system in which his subordinates would reinforce" his preconceived notions by providing him with information they thought he wanted to hear.⁵⁵ Moreover, his megalomania led Saddam to think of himself as a genius strategist whom the Arab masses would see as the successor to Nebuchadnezzar, Saladin, and Egypt's Gamal Abdel Nasser.⁵⁶ This led him to badly misjudge the Arab world's reaction to his invasion. Saddam's officials had assured him that Iraq would face limited opposition from other Persian Gulf monarchies because of their respective tensions with Kuwait. For instance, in a July 1990

---

⁴⁸ Central Intelligence Agency, 2004.

⁴⁹ "Oral History: Tariq Aziz," 1995.

⁵⁰ Holmes, Steven A. "Congress Backs Curbs Against Iraq," *New York Times*, July 28 1990.

⁵¹ Gause, n.d.

⁵² Stein, 1993.

⁵³ Gause, n.d.

⁵⁴ "Oral History: Tariq Aziz," 1995; Central Intelligence Agency, 2004, p. 42.

⁵⁵ Kevin M. Woods and Mark E. Stout, "Saddam's Perceptions and Misperceptions: The Case of 'Desert Storm,'" *Journal of Strategic Studies*, Vol. 33, No. 1, February 2010, p. 36.

⁵⁶ Jerrold M. Post, "The Defining Moment of Saddam's Life: A Political Psychology Perspective on the Leadership and Decision Making of Saddam Hussein During the Gulf Crisis," in Stanley A. Renshon, ed., *The Political Psychology of the Gulf War: Leaders, Publics, and the Process of Conflict*, Pittsburgh: University of Pittsburgh Press, 1993, pp. 49–66.

meeting, the vice chair of the Revolutionary Command Council, Izzat Ibrahim al-Duri, assured Saddam that once Iraq properly explained its grievances against Kuwait,

> Iraq will be excused for any actions it takes by the Arab nation, by the Arab regimes, by the whole world, and by the Arab masses. Not only excused; Iraq will be requested by the Arab masses and Iraqi masses to confront such a conspiracy with all means afforded to Iraq.[57]

U.S. military interviews of captured Iraqi leadership figures suggest that Saddam was genuinely surprised by the backlash of Arab leaders following his invasion.[58] Moreover, Saddam's lack of experience outside the region made him prone to conspiracy theories about a hidden U.S. hand behind everything. He even believed that the United States was behind the overthrow of the shah of Iran.[59]

Despite his poor judgment, Saddam does not appear to have doubted Bush's willingness to respond with force in the event that he invaded Kuwait. Aziz claimed that the Iraqi leadership knew there would be U.S. retaliation; he claimed that they were not given a false impression by Glaspie, and that it "was nonsense to think that the Americans would not attack us" if Iraq invaded Kuwait.[60] Following his occupation of Kuwait, however, Saddam ultimately misjudged Bush's willingness to engage in a long-term fight to free Kuwait. Analogizing from what he knew about the Vietnam War, Saddam believed that public pressure would prevent the United States from maintaining a sustained campaign.[61] According to other Iraqi officials, Aziz also assured Saddam that the United States would take a long time to organize a response to an occupation, which would give Iraq time to strengthen its control over Kuwait.[62] The irony is that some in the Bush administration may have pushed against retaliation had Saddam only pursued a limited invasion. GEN Norman Schwarzkopf, for instance, would have been willing to let Saddam keep small portions of Kuwait. Prior to the Iraqi invasion, when asked about such a potentiality, Schwarzkopf had replied that the United States should not do "a damn thing. The world will not care. It will be a fait accompli."[63]

---

[57] Woods, Pakki, and Stout, 2011, p. 171.

[58] Central Intelligence Agency, 2004.

[59] Woods, Pakki, and Stout, 2011.

[60] "Oral History: Tariq Aziz," 1995.

[61] "Oral History: Wafic Al Samarrai," 1995.

[62] Gause, n.d.

[63] Gordon and Trainor, 1995, 26.

Factors Behind Saddam's Invasion and Implications for Framework

In retrospect, the Bush administration did not employ methods that tend to make for an effective deterrence strategy. Since the Gulf War, it has often been argued that the invasion could have been prevented if the Bush administration had more directly communicated what specific costs the Iraqis would incur if they attacked Kuwait. Perhaps, for instance, the United States should have stationed troops in Kuwait to indicate its capability to push back an Iraqi invasion. Bush could also have issued clearer public statements to stress his commitment to Kuwaiti security.

However, it is important to ask whether Saddam could even have been deterred in the first place. As Saddam so highly valued regime survival, he should have been deterrable at some point in time.[64] While he was willing to incur U.S. retaliation, he was not suicidal. For instance, he refrained from using weapons of mass destruction against U.S. forces because he and his advisers were certain that Washington would respond with nuclear weapons.[65] But by summer 1990—when Saddam's rhetoric and actions led at least a few U.S. intelligence analysts to alert officials of the possibility of an invasion—multiple factors existed that would have posed great obstacles to any attempt at deterrence. While Saddam had been paranoid about the United States as far back as the 1986 Iran-Contra Affair, by 1990 he had become certain that Washington and its allies were trying to overthrow him. Therefore, any attempts to deter Saddam by sending forceful messages could have simply convinced him that an attack was even more imminent. Furthermore, as the economic situation in Iraq worsened, the potential of domestic unrest increased the cost to Saddam of doing nothing—in his eyes, meaning not retaliating against the U.S.-Kuwaiti conspiracy. And once he had mobilized his Republican Guard in Basra on July 21, backing down in the face of U.S. pressure would have seemed even costlier. Perhaps in the early months of 1990, more concerted U.S. attempts to persuade the Kuwaitis to make concessions could have allayed Saddam's fears. But by the summer, any U.S. diplomacy may have been interpreted as an attempt to lull Saddam into complacency. Only a complete Kuwaiti capitulation to Iraqi demands would have staved off an invasion at that point.

At the same time, there is no guarantee that Saddam would not have been provoked to aggression by another alleged conspiracy. According to Aziz, Saddam believed that "Iraq was designated by George Bush for destruction, with or without Kuwait."[66] Therefore, at least by invading Kuwait, Saddam would have a bargaining chip. In the words of Stein, "Once Saddam concluded that the United States was determined to undermine his regime, reassurance and deterrence became virtually impossible, even

---

[64] Post, 1993.

[65] "Oral History: Wafic Al Samarrai," 1995.

[66] "Oral History: Tariq Aziz," 1995.

had the United States clearly defined its commitments and consistently communicated its benign intentions."[67]

Stronger efforts at deterrence may have prevented Saddam from *fully occupying* Kuwait, however. Stationing U.S. forces in Kuwait could have served as a trip wire, preventing the Iraqis from thinking that an occupation would be a fait accompli. After his capture in 2003, Saddam told his American captors that he would not have attacked Kuwait had he realized the level of force with which the United States would respond.[68] Yet this claim contradicts statements made directly after the Gulf War by several senior Iraqi officials and discounts Saddam's heightened level of paranoia at the time. Therefore, it is probably more accurate to say that Saddam would not have occupied Kuwait had he been made aware of the cost.

The U.S. failure to deter Saddam Hussein presents several implications for this report's deterrence framework. Washington was unable to deter Iraq because it lacked a clear understanding of the geopolitical context in the Middle East. Iraq and the United States had been de facto allies in Saddam's war against Iran, but the Iraqi dictator believed the U.S. government to be an enemy intent on his overthrow. In addition, Saddam felt a deep sense of grievance against the Arab monarchies of the Persian Gulf; Iraq had served as a bulwark against Iranian expansion, only for its Arab brothers to betray it. Saddam also viewed himself as a historical leader destined to unite the Arab world under his authority. The United States remained largely ignorant of these realities.

The lack of U.S. understanding regarding Iraq's intentions and regional geopolitics undermined Washington's efforts to shape Saddam's thinking. Positive inducements and relatively vague threats of punishment failed to deter Saddam's occupation of his smaller and much weaker neighbor. The Iraqi regime knew that the Kuwaiti military was no match for its war-hardened military machine. Iran was a weakened regional power while Saudi Arabia was dependent on the United States. Only the United States—the world's only remaining superpower—could prevent an easy Iraqi conquest of Kuwait. The absence of major U.S. forces in the region, a perceived lack of U.S. resolve to defend Kuwait, and the overall mixed—if not confusing—signals from Washington appear to have facilitated Saddam's decision to occupy.

Moreover, Iraq's invasion of Kuwait was not the result of a single decision point, but rather determined by a number of circumstances, including Saddam's belief that he had little choice but to take action against Kuwait in order to survive economically and politically. The authoritarian nature of the Iraqi government and Saddam's paranoid and brutal style of rule only reinforced his decision to invade.

---

[67] Stein, 1993, 135.

[68] Charles A. Duelfer and Stephen Benedict Dyson, "Chronic Misperception and International Conflict: The U.S.-Iraq Experience," *International Security*, Vol. 36, No. 1, Summer 2011, pp. 73–100.

The Aggressor's Intentions and Perceptions

All the framework's variables played a role in convincing Saddam to attack Kuwait. Saddam was unsatisfied with the status quo; he was concerned that lack of adequate oil revenue would lead to domestic unrest. He acted out of a sense of urgency, fearing that the United States and its Arab allies in the Persian Gulf were conspiring to undermine him. Furthermore, an understanding of Iraqi history should have informed analysts of the seriousness of the threat. In the past, Iraq had claimed that Kuwait should be part of its territory. Moreover, Saddam had a history of reckless behavior, exemplified by his invasion of Iran. In retrospect, Saddam's words and actions—threatening statements against Kuwait coupled with amassing his troops on the border—illustrated his high motivation to follow through.

The Status of the Decision Process

Following the Iran-Contra Affair, Saddam had slowly become convinced that the United States was intent on overthrowing him. However, it was not until summer 1990 that he decided there was little alternative other than invading Kuwait. Earlier, he may have still held out hope that the United States would come around to the realization that the two countries were natural allies in the fight against Iran and Islamic extremism.[69]

Conditions for Deterrence Success

The Bush administration failed to send a clear message to Saddam of the costs of invading Kuwait and the benefits of resolving the dispute diplomatically. The Iraqis likely realized the U.S. ability to defeat it. But the lack of clear signaling, including stationing troops in Kuwait to act as a trip wire, left ambiguity as to U.S. willingness to expend force. Furthermore, Saddam decided that fully occupying Kuwait, thereby preventing U.S. troops from amassing there immediately, would buy him time to garner support from the Arab street, as well as allow for U.S. domestic pushback to weaken Bush's resolve. Moreover, Saddam's reading of the U.S. experience in Vietnam, as well as the Reagan administration's decision to withdraw from Lebanon, led him to believe that the American public would not be able to stomach prolonged conflict. Finally, Arab leaders' unwillingness to involve the United States directly in the dispute—including Kuwait's refusal to conduct joint military operations—bolstered Saddam's assessment that his invasion would be a success.

---

[69] Duelfer and Dyson, 2011.

Table C.1
Application of the Framework to the Iraq Case

Category	Variable	Level in Present Case
How motivated was Iraq?	General level of dissatisfaction with status quo and determination to create a new strategic situation.	High. Iraq perceived significant economic and political threats from Kuwait's refusal to comply with Iraqi demands.
	Degree of fear that the strategic situation was about to turn against it in decisive ways.	Mixed. No immediate threat to territory or regime, but Kuwaiti intransigence perceived as possible basis for larger turn in strategic context.
	Level of national interest involved in specific territory of concern.	Perceived high, in relation to Saddam's ambitions. In objective terms, arguably mixed to low.
	Urgent sense of desperation or requirement to act; whether aggressor is locked into course of action.	High. Perceived rather than real, but Saddam got himself to a place where he felt he could not simply stand by.
	Degree of aggressive, reckless, risk-accepting opportunism.	High. Saddam engaged in significant risk-taking.
	Level of motivated reasoning in play; degree of wishful thinking, misperception of basic strategic context.	High. Closed decision process; Saddam's megalomania led to dangerous levels of motivated reasoning.
Was the United States clear and explicit regarding what it sought to prevent and what actions it would take in response?	Precision in the type of aggression the United States sought to prevent.	Low. Washington never made an explicit commitment to Kuwait's defense.
	Clarity in the actions that would be taken in the event of aggression.	Low. Very weak statements of likely response.
	Forceful communication of these messages to outside audiences, especially potential aggressor(s).	Low. Few public statements, no effort to ensure Saddam got the message.
	Timely response to warning with clarification of interests, threats.	Low. As buildup continued, the United States still did not clarify intentions.
Did Iraq view U.S. threats as credible and intimidating?	Actual and perceived strength of the local military capability to deny the presumed objectives of the aggression.	Low. U.S. forces in the region were insufficient to deny gains.
	Degree of automaticity of U.S. response, including escalation to larger conflict.	Low. No military commitments in place and no local trip-wire forces to guarantee U.S. response.
	Degree of actual and perceived credibility of political commitment to fulfill deterrent threats.	Low. No public U.S. stance.
	Degree of national interests engaged in state to be protected.	Mixed to high. Oil considerations were significant but could theoretically be met by continued Iraqi production.
	Reputation for resolve with potential aggressor.	Low. U.S. had been courting Saddam, who was viewed as counterweight to Iran.
	Degree of threat posed to attacker's values and interests by the specific responses threatened by the defender.	Low. No specific responses threatened.

APPENDIX D

Qualitative Case Study Analyses: NATO's Northern Flank in the Cold War

During the Cold War, NATO's Northern Flank comprised Denmark and Norway; it was lightly defended, removed from the Central Front, and vulnerable to potential military aggression from the east.[1] In case of an attack, the two countries' militaries would have had difficulty holding out until allied reinforcements arrived. The governments of Denmark and Norway further imposed severe limitations on their participation in the NATO alliance by banning foreign military personnel, bases, and nuclear weapons from their territories. The overwhelming Soviet military presence just across the USSR's shared border with Norway—comprising naval, air, ground, and nuclear forces—compounded the situation. The United States and NATO military planners were for a large portion of the Cold War much more focused on the situation along the Central Front and tended not to consider the Northern Flank a priority. This confluence of factors made the Northern Flank a seemingly obvious target for Soviet expansion and made Soviet military victory in case of a conflict seem almost inevitable. Yet that did not occur. The reasons for this have potentially important implications for U.S. extended deterrence in other situations.

The situation of the Northern Flank countries throughout the entire period of the Cold War presents a unique case of U.S. extended deterrence. The United States did not begin to invest serious efforts into deterring the Soviet Union in this region until the last decade of the Cold War, but the Soviets did not undertake military aggression against Denmark or Norway in the meantime. U.S. deterrence efforts here ultimately succeeded due to a combination of the Soviets' own limited objectives for the region, the clarity of the U.S. deterrence messaging, and the aggressiveness of U.S. and NATO deterrence.

---

[1] Certain sources also include the northern German state of Schleswig-Holstein as part of the Northern Flank. See, for example, Ragnhild Sohlberg, *Analysis of Ground Force Structures on NATO's Northern Flank*, Santa Monica, Calif.: RAND Corporation, N-1315-MRAL, 1980, p. 3. However, because defense of Schleswig-Holstein also had implications for the defense of West Germany, that is something beyond the scope of this appendix. The present appendix instead focuses only on Denmark and Norway because both are Nordic member states of NATO located in continental Europe and, moreover, faced many of the same challenges during the Cold War.

This appendix is organized as follows: the first section provides background to U.S. extended deterrence of the Soviet Union along the Northern Flank. It looks particularly at the strategic value of Denmark and Norway to both the United States and the Soviet Union. It then provides an overview of the security postures of the various actors and the overall security situation in the region. The second section looks at U.S. extended deterrence through application of the framework articulated in Chapter Three of this report. It examines the potential aggressor's motivations, the clarity of the defender's message, and the credibility of the defender's threat. It then draws lessons from the case. The third section concludes the appendix and draws together key findings.

Background

The Strategic Value of the Northern Flank Countries

The United States and Soviet Union quickly recognized the strategic value of the Northern Flank countries. In a 1948 report to the president, the National Security Council made the following assessment of the region:

> The Scandinavian nations are strategically important both to the United States and the USSR. They lie astride the great circle air route between North America and the strategic heart of Western Russia, are midway on the air route between London and Moscow, and are in a position to control the exits from the Baltic and Barents seas. Domination of Scandinavia would provide the Soviets with advanced air, guided-missile and submarine bases, thus enabling them to advance their bomb line to the west, to threaten allied operations in the North Atlantic, and to form a protective shield against allied sea or air attack from the Northwest.[2]

Moreover, just across the 120-mile border shared with Norway were the Soviet Union's "only two good-sized ports that remain[ed] open all through the year," Murmansk and Pechenga, making the security of these ports and the adjacent areas vital to Soviet naval operations.[3] Farther out from the continent was the Greenland-Iceland-UK (GIUK) gap (see Figure D.1). Control of this gap, along with control of the Baltic straits and the Norwegian littoral, was "critical to the superpowers' strategic nuclear balance." The GIUK gap itself served as the "first line of defense for antisubmarine warfare (ASW) against Soviet nuclear-powered ballistic missile submarines (SSBNs)

---

[2] National Security Council, *NSC 28/1: A Report to the President by the National Security Council on the Position of the United States with Respect to Scandinavia*, Washington D.C.: National Security Council, September 3, 1948b, pp. 1–2.

[3] Nils Ørvik, "Soviet Approaches on NATO's Northern Flank," *International Journal*, Vol. 20, No. 1, Winter 1964–1965, p. 55.

Figure D.1
The GIUK Gap and the Strategic Situation of the Northern Flank

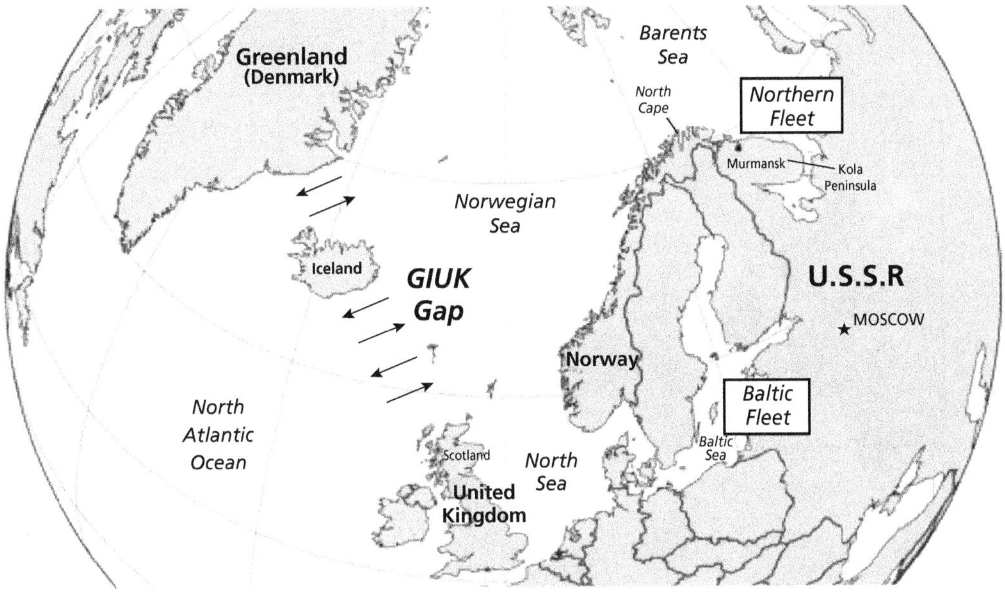

SOURCE: CDRSalamander, "Once More unto the Gap."

leaving the Baltic and Barents seas."[4] This made the GIUK gap both a major asset to NATO and a potentially dangerous choke point for the Soviet Union.

Context: The Strategic Situation Along the Northern Flank

The recognized strategic value of the Northern Flank countries notwithstanding, for most of the Cold War this region was not the site of major contestation between U.S.-NATO and Soviet–Warsaw Pact forces. In fact, as Figure D.2 makes clear, the United States kept very low levels of forces in the region. A confluence of factors ensured this stability—most notably, the perceptions that the various players had of the regional security situation. Finland signed a Treaty of Friendship, Cooperation and Mutual Assistance with the USSR on April 6, 1948, thereby guaranteeing its own neutrality and its status as a reliable buffer state for the Soviets.[5] Sweden, having managed to avoid occupation during World War II, decided to maintain armed neutrality in the postwar period, creating yet another buffer between the East and the West.

Both Denmark and Norway decided to abandon their long-standing traditions of neutrality and join NATO because of the experience of being invaded and occupied

---

[4] Manfred R. Hamm, *Ten Steps to Counter Moscow's Threat to Northern Europe*, Backgrounder No. 356, Washington, D.C.: Heritage Foundation, May 30, 1984, pp. 2–3.

[5] John Lukacs, "Finland Vindicated," *Foreign Affairs*, Fall 1992.

Figure D.2
U.S. Military Personnel Permanently Stationed in Northern Europe, 1950–1992

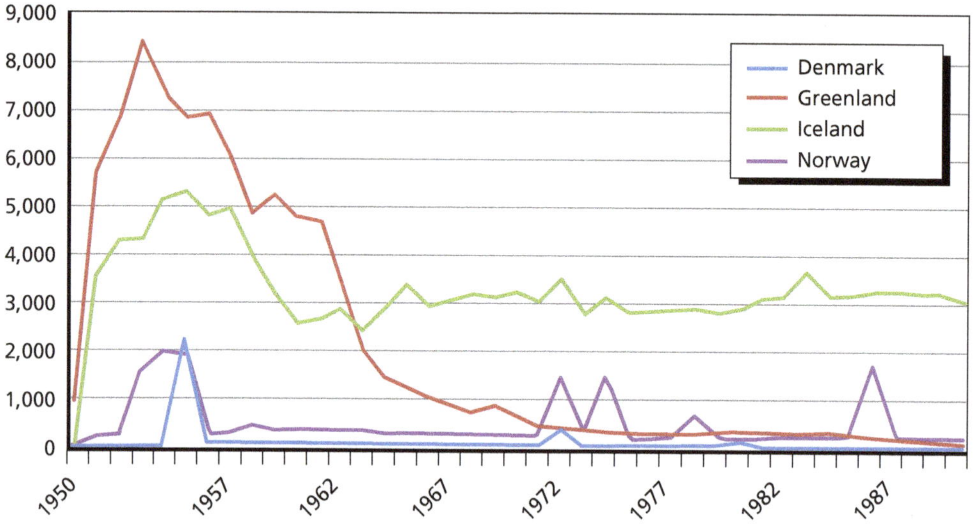

SOURCE: Defense Manpower Data Center, "Worldwide Manpower Distribution by Geographical Area (M05): Historical Reports—Military Only, 1950, 1953–1999," n.d.

by Nazi Germany. At the same time, however, they remained wary of antagonizing the Soviet Union and so pursued a strategy of simultaneous deterrence and reassurance. Accordingly, they joined NATO to serve as an ultimate security guarantee, but they simultaneously imposed a number of important restrictions on their participation in the Alliance.

Denmark, for example, banned the establishment of foreign military bases on its territory during negotiations to join NATO. In the early 1950s, Denmark forbade the stationing of foreign forces on its territory.[6] From the late 1950s onward, Denmark also had an official policy that "excluded all nuclear weapons from Danish territory, including Greenland."[7] Around 1988 Denmark became involved in a dispute with the United States about whether to permit NATO ships carrying nuclear weapons to visit its ports. This mirrored a U.S. dispute with New Zealand from two years earlier that ultimately saw the United States suspend its security obligations to New Zealand

---

[6] Carsten Holbraad, "Denmark: Half-Hearted Partner," in Nils Ørvik, ed., *Semialignment and Western Security*, New York: St. Martin's Press, 1986, p. 19. This ban, however, did not extend to the Danish dependency of Greenland, which was used to station thousands of U.S. military personnel from the 1950s until the late 1960s, when the number of personnel dipped below a thousand for the first time.

[7] Holbraad, 1986, p. 19; Unofficially, however, Danish prime minister and foreign minister H. C. Hansen had actually given the go-ahead for nuclear weapons to be stationed in Greenland. Details of this informal agreement did not emerge publicly until 1995. For more details, see Nikolaj Petersen, "SAC at Thule: Greenland in the U.S. Polar Strategy," *Journal of Cold War Studies*, Vol. 13, No. 2, Spring 2011, pp. 104–105.

as set out in the Australia, New Zealand, United States Security Treaty.[8] The United States and Denmark defused tensions only after coming to an understanding whereby "NATO ships [would] be able to visit Danish ports without confirming or denying whether they carry nuclear weapons."[9]

For Norway, Allied Forces Northern Europe (AFNORTH) had its headquarters in Kolsås and was "responsible for the defense of Norway, Denmark, Schleswig-Holstein and the Baltic approaches." AFNORTH included

> one Norwegian and one Danish division, as well as the tactical air forces of these countries and their naval forces. One German division in Schleswig-Holstein and units of the German Baltic fleet [were] also committed to AFNORTH.[10]

Kolsås, located some eight miles west of Oslo, is distant from Norway's border with the Soviet Union and is roughly 860 miles from Finnmark, Norway's extreme northeast county. In contrast, the Russian city of Murmansk, on the Kola Peninsula, is roughly 200 miles away from Finnmark. In case of a crisis, Soviet forces stationed on the Kola Peninsula would be able to cross over into northern Norway far more quickly than AFNORTH Norwegian and Danish forces—and NATO reinforcements dispatched from elsewhere—could reach the border. (See Figure D.3.)

Norway forbade any permanent stationing of NATO troops on its soil, banned military maneuvers near the Soviet border, and refused to store nuclear or chemical weapons on its territory, with the additional stipulation that all vessels making port calls had to be nuclear-free.[11] By these measures, Denmark and Norway believed that the Soviet Union would be deterred from aggression, but they were also reassured that their countries would not be used as staging grounds for an allied attack on Soviet territory.

The United States appeared most interested in Denmark, primarily to maintain its preexisting bases in Greenland. It otherwise had limited interest in the region as a whole.[12] The United States signed an agreement with Denmark in 1941 that allowed it to build military facilities on Greenland, and in 1951 it moved to sign another that would allow it continued access to Greenland.[13] In various reports, the National

---

[8] Bureau of East Asian and Pacific Affairs, "U.S. Relations with New Zealand," U.S. Department of State, February 14, 2017.

[9] Michael Gordon, "Denmark Agrees to Nuclear Policy," *New York Times*, June 8, 1988, p. A14.

[10] "North Atlantic Treaty Organization," *Military Balance*, Vol. 61, No. 1, 1961, p. 10.

[11] "Tensions Rise on Norwegian-Soviet Border," *New York Times*, December 7, 1986.

[12] Holbraad, 1986, pp. 17–18.

[13] U.S. Department of State, *Defense of Greenland*, April 9, 1941; "Defense of Greenland: Agreement Between the United States and the Kingdom of Denmark, April 27, 1951," Avalon Project, Lillian Goldman Law Library, Yale Law School, April 27, 1951.

Figure D.3
The Position of Kolsås and Murmansk Relative to Finnmark

SOURCE: Google Earth.
RAND RR2451A-D.3

Security Council also stressed that Greenland's "use by an enemy must be denied" and that it "must continue to be available to the United States for military purposes" because of the air bases and early warning installations located there.[14] Overall, U.S. and NATO planners focused most on the inner German border, where they believed conflict with the Warsaw Pact nations most likely to break out. The Nordic countries served simply as a "'tactical northern flank' to the Central Front."[15] Thus, the two sides did not hotly contest the Northern Flank for much of the Cold War.

For the Soviet Union it was especially important that the Nordic countries not become a springboard for a NATO attack.[16] According to Russian historian Mikhail Suprun, the Soviet Union at least partially achieved this aim in 1945. That year, the

---

[14] National Security Council, *NSC 28: A Report to the National Security Council by the Executive Secretary on the Position of the United States with Respect to Scandinavia*, Washington, D.C.: National Security Council, August 26, 1948a, p. 4; National Security Council, *NSC 6006/1: U.S. Policy Toward Scandinavia (Denmark, Norway and Sweden)*, Washington, D.C.: National Security Council, April 6, 1960, p. 3.

[15] Gjert Lage Dyndal, "50 Years Ago: The Origins of NATO Concerns About the Threat of Russian Strategic Nuclear Submarines," *NATO Review*, March 24, 2017.

[16] Holbraad, 1986, pp. 17–18.

Western Allies gave secret guarantees to Joseph Stalin that they would "refrain from establishing military bases in Finnmark, a Norwegian county bordering Soviet territory," which in turn prompted the Soviet Union to withdraw its troops from Norway and drop its demands for concessions in the country's northern region and Svalbard.[17] Formerly known as Spitsbergen, Svalbard is an archipelago located roughly 450 miles north of Norway in the Arctic; its proximity to both Greenland and the northern coast of Asia made it of "commanding importance to Russia's northern sea route, which extends from Archangel to the Bering Straits" at the time.[18] The February 1920 signing of the Svalbard Treaty by Denmark, France, Italy, Japan, the Netherlands, Norway, Sweden, the United Kingdom and its dominions, and the United States granted Norway sovereignty of Svalbard. The Soviet Union approved the treaty in 1924 and reaffirmed it without reservations in 1935. In 1944, the Soviet Union requested that Norway revise the treaty, but Norway argued it could not without all the other signatories. When the Soviet Union made another request in 1947, Norway offered to discuss only economic readjustments, such as in taxes and improvement of facilities, but not sovereignty. The Soviet Union subsequently dropped the issue.[19]

Secret guarantees aside, Soviet leaders did not remain complacent regarding the region. The Soviet Union continued to build up its military strength along the border with Norway throughout the Cold War. Norwegian historian Nils Ørvik notes that Norway practiced "a strategy of deliberate military weakness" in order to reassure the Soviets, but "in concrete terms, measured area for area, the Norwegian [confidence-building measures] policy [was] a one-way street."[20] By the 1980s the Soviet Union had built up significant ground, naval, air, and missile forces across the border with Norway. In peacetime the Kola Peninsula hosted more than 225 air defense fighters, many smaller ones, land-based bombers, fighters of Soviet naval aviation, more than 100 surface-to-air missile installations, and numerous radar tracking sites.[21] Additionally, by 1980 the Soviets had stationed eight motorized rifle divisions (MRDs), one airborne division, and one artillery division in the Leningrad Military District.[22] By 1989 this presence increased to 11 MRDs, two of which were mobile, one airborne division, one artillery brigade, four Scud brigades, and one air assault brigade.[23] The Soviet Northern Fleet was also stationed on the Kola Peninsula and contained "nearly

---

[17] Quoted in Alexey Golubev, "*Kholodnaya voina v Arktike* (Review)," *Journal of Cold War Studies*, Vol. 14, No. 1, Winter 2012, p. 144.

[18] John J. Teal, Jr., "Europe's Northernmost Frontier," *Foreign Affairs*, January 1951, pp. 263–264.

[19] Teal, 1951, pp. 271–272.

[20] Nils Ørvik, ed., *Semialignment and Western Security*, New York: St. Martin's Press, 1986, pp. 210–211.

[21] Richard D. Hooker, Jr., "NATO's Northern Flank: A Critique of the Maritime Strategy," *Parameters*, June 1989, p. 31.

[22] "The United States and the Soviet Union," *Military Balance*, Vol. 81, No. 1, 1981, p. 12.

[23] "The Soviet Union," *Military Balance*, Vol. 90, No. 1, 1990, p. 39.

two-thirds of the Soviet Union's retaliatory submarine-launched nuclear strike force, as well as major surface units."[24] This nuclear strike force could be used for long-range strategic deterrence such as by firing submarine-launched ballistic missiles over the North Pole.

U.S. Attempts to Deter the USSR from Using Force Against the Northern Flank

The imbalance of forces along the Northern Flank (and particularly along the Norwegian-Soviet border), coupled with the strategic importance of this region to both the United States and the Soviet Union, did not ultimately lead to Soviet military action or an armed clash between the two sides. There were moments of alarm—for example, in 1979 when the Soviet invasion of Afghanistan and internal exile of Nobel Peace Prize winner Andrei Sakharov provoked strong reactions among the Norwegian public.[25] One contemporary analyst at the Strategic Studies Institute noted that the Soviet Union might be tempted to attack Norway in order to seize its eight major airfields and its harbors. Controlling the airfields would have allowed the Soviets to significantly extend the range of its weapons, such as the Backfire bomber, while possession of the easily defended harbors would have allowed them to "significantly increase the reaction time for their ships to enter the North Atlantic Ocean and decrease the length of their sea logistics lines to supply these ships."[26]

The same analyst, meanwhile, argued that Denmark held strategic importance "astride the narrow Danish straits" and was "particularly vulnerable to amphibious assault." However, the Soviets may have felt wary about moving against Denmark, as "an attack on Denmark might be interpreted as an immediate threat to 'mainland' Central Europe and could signal Soviet intentions to expand their northern thrust beyond the immediate objective of securing only the Northwestern Region."[27] A 1989 CIA research paper, however, assessed that "in the event of a NATO–Warsaw Pact war, Pact planners remain committed to a coordinated, phased offensive operations against Denmark. . . . Pact plans appear to envisage initiating the operation through Jutland into Denmark prior to NATO reinforcement."[28]

---

[24] Robert K. German, "Norway and the Bear: Soviet Coercive Diplomacy and Norwegian Security Policy," *International Security*, Vol. 7, No. 2, Fall 1982, p. 56.

[25] Leonard Downie, Jr., "Scandinavia, Alarmed by Afghanistan, Reviews Its Defenses," *Washington Post*, February 5, 1980.

[26] William K. Sullivan, *Soviet Strategy and NATO's Northern Flank*, Carlisle, Penn.: Strategic Studies Institute, U.S. Army War College, May 1, 1978, pp. 8–9.

[27] Sullivan, 1978, pp. 9–10.

[28] Central Intelligence Agency Directorate of Intelligence, *Warsaw Pact: Planning for Operations Against Denmark*, April 1989, p. iii.

Despite these plausible motives and opportunities for an attack, it never materialized. U.S. deterrence in this case appears to have succeeded first because the Soviet Union held limited objectives, and second, because a gradual shift in U.S. and NATO strategic thinking about northern Europe from the late 1960s onward caused the United States and its allies to become more assertive and proactive in their defense of the region.

The Aggressor's Motivation

One window into the Soviet Union's objectives in the region is the Nordic Balance theory, which holds that the region existed in a state of balance because of the different security postures adopted by the four continental Nordic countries: Denmark and Norway's membership in NATO (but with important reassuring caveats), Sweden's armed neutrality, and Finland's special relationship with the Soviet Union. This balance was self-correcting: If the Soviet Union attempted to pressure Finland and/or Sweden in some way, then that would lead Denmark and Norway to review their restrictions on their participation in NATO and vice versa. Mindful of such possibilities, the two superpowers decided to keep their involvement in the region limited.[29] The extent to which the Nordic Balance existed in practice was controversial. In 1963, Ørvik pointed out that while the balance might have existed in 1949, by 1963 it had "become quite irrelevant in view of the enormous build-up in Soviet military strength in the Nordic Cap area."[30] That is, "the theoretical Norwegian response to a Soviet move, namely, to invite foreign military assistance (essential if the balance theory is to be considered valid), was no longer credible, since the no bases-policy laid down in 1949 had crystallized into dogma."[31] Robert German, writing in 1982, echoed this argument and further noted that the Nordic Balance concept was not widely accepted outside Denmark and Norway: "Swedish officials rarely refer to it and Finns almost never."[32]

The lack of a Nordic Balance, however, did not automatically translate into a Soviet desire to seize control of the region. Such a desire might have existed in Joseph Stalin's time, when official Soviet doctrine held that there could only be "two camps," with no room for neutrality. Indeed, Stalin "voiced doubts that small states would be able to maintain neutrality." Under Nikita Khrushchev, however, "peaceful coexistence" became official Soviet policy, while the Nordic countries (particularly Finland and Sweden) became a showcase of such a coexistence.[33] Khrushchev and subsequent

---

[29] German, 1982, pp. 77–78.

[30] George Schöpflin, "NATO and the Nordic Balance," *World Today*, Vol. 22, No. 3, March 1966, p. 117.

[31] Schöpflin, 1966, p. 118.

[32] German, 1982, p. 78.

[33] Wolfgang Mueller, "The USSR and Permanent Neutrality in the Cold War," *Journal of Cold War Studies*, Vol. 18, No. 4, Fall 2016, pp. 148–179.

Soviet leaders saw advocating neutrality as a pragmatic way to "keep independent countries out of Westerns alliances or to lure them away." In northern Europe, advocating neutrality and "peaceful coexistence" helped ensure an effective buffer between Soviet and U.S. forces.

Within the general framework of "peaceful coexistence," the Soviet Union's major concerns were (1) that the Nordic countries—especially Norway as the only NATO member to share a border with the Soviet Union—not be used as the staging ground for an allied attack, and (2) that allied ships and other military assets be kept as far away possible from nuclear forces on the Kola Peninsula. This defensive effort came to be known as the Bastion Concept, which was "centered on defending and securing the Soviet sea-based nuclear forces located in the vicinity of the Kola Peninsula."[34] The Soviet Northern Fleet defined the Barents Sea as a closed area for its SSBNs—effectively, a "bastion." This bastion "became heavily defended by attack submarines, surface vessels and air power," all of which was done in order to ensure the survivability of the Soviets' second strike capability.[35] In wartime, military invasion of northern Norway could achieve the Soviets' twin goals (of preventing the Northern Flank countries from being used as a springboard and of keeping NATO forces away from the Kola bastion), as U.S. military analysts of the time noted.[36] In peacetime, however, the Soviets could pursue these two goals via subtler means, with less risk of a military confrontation with NATO.

In practice this meant applying a combination of political and military bullying and occasionally other inducements to try to isolate Denmark and Norway from the Alliance. The military buildup along the Norwegian-Soviet border was one way of achieving this and other goals, such as securing the safety of the bastion. Another way was to protest vigorously and lecture Denmark and Norway about their obligations whenever either deviated from their self-imposed policies of prohibiting allied forces or equipment on their territory. For example, in 1951 the Soviets sent a strongly worded note to the Norwegian government to protest the agreement to locate AFNORTH headquarters in Kolsås. In 1957, "when NATO was considering the American proposal to establish nuclear stockpiles and deploy intermediate-range ballistic missiles in Europe," Soviet premier Nikolai Bulganin wrote to warn the Norwegian and Danish governments against offering up their territory for such use. He noted that "NATO bases in their countries would constitute 'legitimate targets' for Soviet hydrogen bombs." Finally, in 1959, the Soviet ambassador in Oslo sent a demarche to warn

---

[34] John Andreas Olsen, "Introduction: The Quest for Maritime Supremacy," in John Andreas Olsen, ed., *NATO and the North Atlantic: Revitalising Collective Defense*, Whitehall Paper No. 87, Abingdon, England: Routledge, 2016, pp. 3–4.

[35] Dyndal, 2017.

[36] Robert F. Kernan, *Norway and the Northern Front: Wartime Prospects*, Maxwell Air Force Base, Ala.: Air War College, May 1989.

the Norwegian government against establishing supply depots for the German Navy operating under the auspices of the Baltic Approaches Command, itself subordinate to AFNORTH.[37]

Although such constant pressure could have had the effect of hardening attitudes toward the Soviet Union, on the Norwegian side there was genuine concern that it might eventually cause the population to simply give in and become (even) more accommodating toward Soviet policy.[38] American analysts feared as much. An April 1960 National Security Council (NSC) report argued that the Nordic countries were "susceptible to the idea of disengagement, thinning out of forces and negotiated settlement in continental Europe."[39]

At various times, Soviet leaders either supported or directly made proposals for transforming the whole of northern Europe into a nuclear-weapons-free zone (NWFZ). This position enjoyed domestic popularity in Denmark and Norway, but other NATO governments viewed it as having the potential to do severe damage to the Alliance.[40] In June 1959 Soviet leader Nikita Khrushchev proposed a "nuclear and missile-free zone in the Scandinavian peninsula and Baltic area."[41] In 1981, Leonid Brezhnev even hinted that the Kola Peninsula and Murmansk could be included in an eventual NWFZ (although official Soviet news sources subsequently explicitly rejected this).[42] His successor Yuri Andropov reiterated the proposal in June 1983 during a visit by Finnish president Mauno Koivisto, this time offering to extend the zone to include the Baltic Sea.[43] Regarding these repeated overtures, in written responses given to the Swedish newspaper *Svenska Dagbladet* in 1987, President Ronald Reagan noted, "I do not think that a northern European nuclear-free zone would increase security in Scandinavia. . . . International agreements that would appear to create two categories of NATO members would weaken the alliance and thereby increase instability and undercut deterrence."[44]

---

[37] German, 1982, pp. 62–64.

[38] This concern was expressed by Norwegian Major General Gunnar Helset, who served with AFNORTH around 1985, in Rolf Soderlind, "Norwegian Airfield Crucial in World War III Scenario," UPI, April 11, 1985.

[39] National Security Council, 1960.

[40] John Vinocur, "Brandt's Soviet Visit Troubles Schmidt and NATO," *New York Times*, July 13, 1981.

[41] Walter C. Clemens and Franklyn Griffiths, "The Soviet Position on Arms Control and Disarmament—Negotiation and Propaganda, 1954–1964," Cambridge, Mass.: Center for International Studies, Massachusetts Institute of Technology, February 1, 1965, p. 52.

[42] Chris Mosey, "Nordic Enthusiasm for Nuclear-Free Zone Pleases USSR, but Not US," *Christian Science Monitor*, July 3, 1981; "Nordic Nuclear Weapon–Free Zone," House of Lords Debate, Vol. 424 cc895-7, October 27, 1981.

[43] John F. Burns, "Andropov Offers Atom-Free Baltic," *New York Times*, June 7, 1983.

[44] Ronald Reagan, "Written Responses to Questions Submitted by the Swedish Newspaper Svenska Dagbladet," September 22, 1987, American Presidency Project.

Robert German notes that

> throughout the post-war years Moscow has sought to weaken Scandinavian ties with the West and to make of Northern Europe a sort of neutral, ideally pro-Soviet, extension of the buffer zone which it created by force in the Baltic Republics and Eastern Europe.[45]

Other contemporary analyses offered similar views. Nils Ørvik, for instance, argued that by the 1960s NATO was already experiencing a number of internal cohesion problems. In such an atmosphere, an attack on Denmark and Norway would have had the undesired effect of strengthening NATO cohesion. This risk made indirect control through political means the better strategy for Soviet leaders.[46] Seizing control of these two countries, moreover, would have resulted in the loss of a useful buffer. Arne Olav Brundtland argues that "the interest of both the Soviet Union and the United States in the Finno-Scandinavian area during the post–World War II period . . . has been one of denial, not of possession."[47] Örjan Berner, the Swedish ambassador to Moscow from 1989 to 1994, also suggested that the Soviet attitude during the postwar period had been one of "positive vigilance combined with steady pressure in order to maintain and, if possible, improve a fairly satisfactory situation and to insure against any negative developments arising from internal or external forces."[48] That is, converting Denmark and Norway into officially neutral states and extending the northern European buffer would have been ideal but was not worth the risk of general war with NATO and the United States. Furthermore, the security situation was otherwise tolerable, since the two countries had already adopted caveats on NATO activities that reassured the Soviet leadership.

The Clarity of the Defender's Message

The U.S. message regarding Denmark and Norway cannot be separated from its message regarding NATO as a whole. As members of the Alliance, these two countries benefited from the U.S. commitment to defend Western Europe from potential Soviet and Warsaw Pact aggression in the same way that countries like West Germany or France did. These benefits endured even when U.S. administrations rarely spoke specifically about Denmark or Norway. A review of their speeches throughout

---

[45] German, 1982, p. 55.

[46] Ørvik, 1964–1965.

[47] Arne Olav Brundtland, "The Nordic Balance: Past and Present," *Cooperation and Conflict*, Vol. 1, No. 4, 1966, p. 30.

[48] Örjan Berner, *Soviet Policy Toward the Nordic Countries*, New York: University Press of America, 1986, p. 108, quoted in Robert W. Janes, "The Soviet Union and Northern Europe: New Thinking and Old Security Constraints," *Annals of the American Academy of Political and Social Science*, Vol. 512, No. 1, November 1990, p. 166.

the decades shows that successive U.S. presidents clearly messaged the U.S. commitment to defending Europe. In July 1963, for example, President John F. Kennedy stated, "The NATO treaty pledges us all to the common defense—to regard an attack on one as an attack on all, and respond with all the force required—and that pledge is as strong and unshakable now as it was the day it was made."[49] In 1964 his successor, President Lyndon Johnson, echoed this sentiment when he spoke to the NATO Parliamentarians Conference: "The United States has made certain commitments both real and substantial, and we will meet them all. Let no one, ally or adversary, ever doubt America's determination to fulfill its role in the alliance, to live up to its obligations."[50]

On the Republican side, on the occasion of the twentieth anniversary of NATO in 1969, President Richard Nixon told the North Atlantic Council that "NATO is needed; and the American commitment to NATO will remain in force and it will remain strong. We in America continue to consider Europe's security to be our own."[51] However, by the early 1970s, European confidence in U.S. commitment began to wane. As then–Assistant Secretary of State for European Affairs Martin Hillenbrand noted in a memorandum to then–Secretary of State William Rogers in 1971, "From the top of Norway to the tip of Italy there is a growing conviction that the United States will disengage from Europe; the only question is when."[52] In the midst of this uncertainty, President Gerald R. Ford gave one of the most forceful declarations of U.S. commitment to the defense of Europe yet. In May 1975, in a speech to the North Atlantic Council on the occasion of the twenty-fifth anniversary of NATO, Ford stated, "The United States of America unconditionally and unequivocally remains true to the commitments undertaken when we signed the North Atlantic Treaty, including the obligation in Article V to come to the assistance of any NATO nation subjected to armed attack."[53] Ten years later, in remarks prepared for the occasion of the thirty-fifth anniversary of NATO, President Ronald Reagan stated, "And I hope that the Soviet leadership will finally realize it is pointless to continue its efforts to divide the alliance. We will not be split. We will not be

---

[49] John F. Kennedy, "Remarks in Naples at NATO Headquarters," July 2, 1963, American Presidency Project.

[50] Lyndon B. Johnson, "Remarks to Members of the NATO Parliamentarians Conference," September 18, 1964, American Presidency Project.

[51] Richard Nixon, "Address at the Commemorative Session of the North Atlantic Council," April 10, 1969, American Presidency Project.

[52] "Information Memorandum from the Assistant Secretary of State for European Affairs (Hillenbrand) to Secretary of State Rogers," in James E. Miller and Laurie Van Hook, eds., *Foreign Relations of the United States, 1969–1976*: Vol. XLI, *Western Europe; NATO, 1969–1972*, Washington, D.C.: GPO, 2012, p. 323.

[53] Gerald Ford, "Text of an Address Before the Council of the North Atlantic Treaty Organization in Brussels," May 29, 1975, American Presidency Project.

intimidated. The West will defend democracy and individual liberty. And the West will protect the peace."[54]

In addition to public commitments to NATO, senior U.S. officials occasionally addressed Denmark and Norway specifically. In 1952, President Harry S. Truman wrote letters to the chairmen of several Senate committees, arguing in favor of continuing financial aid to Denmark. To do otherwise, he noted, "would clearly be detrimental to the security of the United States by weakening the defense of NATO, contributing to the strength of the Soviet Union, fostering the political and propaganda objectives of the communist bloc."[55] In 1962, during a visit from Norwegian prime minister Einar Gerhardsen, President Kennedy and Gerhardsen released a joint statement in which they "reaffirmed their determination to give unstinting support to the NATO Alliance. It is imperative, they recognized, for the West to maintain a position of strength and to stand fast in face of outside provocations or pressures."[56] In January 1979, during a meeting with Deng Xiaoping, then the vice premier of China, Secretary of Defense Harold Brown stated that "if the Soviets attempt[ed] to attack [the Northern Flank] NATO [had] plans to reinforce very quickly by both air and sea."[57]

The United States was therefore clear in its deterrence messaging. Successive presidents of the United States, whether Democrat or Republican, publicly and forcefully reiterated the U.S. commitment to NATO and the defense of Western Europe. Through strategies like massive retaliation and flexible response, U.S. presidents further clarified how they would respond to Soviet aggression against Western Europe. These commitments to Western Europe and threats against the Soviet Union naturally extended to Denmark and Norway. As shown in the next section, concrete U.S. actions further served to reinforce the credibility of U.S. extended deterrence.

The Credibility of the Defender's Threats
Gjert Lage Dyndal, the former dean of the Norwegian Air Force Academy, argues that the perception NATO planners had of the "High North"—the area surrounding where the Norwegian and Soviet borders met—began to shift in the late 1960s because of several key developments. In late 1967, NATO officially adopted the U.S. strategy of flexible response as Military Committee document 14/3 (MC 14/3), providing the Alliance with response options beyond massive retaliation. Options included direct

---

[54] Ronald Reagan, "Remarks on the 35th Anniversary of the North Atlantic Alliance," May 31, 1984, American Presidency Project.

[55] Harry S. Truman, "Letter to Committee Chairmen on the Need for Continuing Aid to Denmark," July 25, 1952, American Presidency Project.

[56] John F. Kennedy, "Joint Statement Following Discussions with Prime Minister Gerhardsen of Norway," May 11, 1962, American Presidency Project.

[57] "Memorandum of Conversation," January 29, 1979, in Adam M. Howard, ed., *Foreign Relations of the United States, 1977–1980*: Vol. XIII, *China*, Washington, D.C.: GPO, 2013, p. 747.

defense, deliberate escalation (such as by opening up another front or attacking at sea), and general nuclear response.[58] Around the same time, the Supreme Allied Commander Atlantic became concerned about an increase in Soviet naval activity, which led to the publication of two studies on maritime strategy in 1965 and 1967. Additionally, starting in 1967, NATO planners also became increasingly aware of the Soviet SSBN threat. In 1968 they instituted the concept of "External Reinforcement of the Flanks," which contained four operational elements: "the Allied Mobile Force (AMF), the Standing Naval Force, the Quick Reaction Mobile Force (QRMF) and the Maritime Contingency Force," which were "divided into two sets of forces, the Immediate Reaction Force (the AMF and the Standing Naval force [sic]), and the Reinforcement Forces (the QRMF and the Maritime Contingency Force)."[59] The following year they produced the Brosio study (named after the NATO secretary general, Manlio Brosio), which looked at both the relative naval strength and maritime doctrines of NATO and the Soviet Union.[60] This series of developments made NATO planners increasingly aware of the strategic value of the Nordic countries—Norway, in particular—and the potential military threat that the Soviet Union posed there. An increased presence did not materialize until later in the 1970s, however, because of the "general decline of most Western forces during the late 1960s . . . due to the U.S. withdrawal from Vietnam and the general trend of détente in the West. From the mid-1970s, however, the Western naval presence in the High North steadily increased."[61]

From the mid-1970s onward, U.S. and NATO military planners adopted other measures to strengthen their presence on the Northern Flank. In 1975, SACEUR GEN Alexander Haig introduced a new NATO exercise program called Autumn Forge. This effort linked previously autonomous multinational and NATO exercises and extended geographically from Norway to Turkey and took place from September through November.[62] The component exercises linked directly to the Nordic countries included Bold Guard and Northern Wedding. The former took place in Denmark and

---

[58] Flexible response was an initiative of the Kennedy administration that replaced the Eisenhower administration's strategy of massive retaliation. For more on flexible response within the American context, see, Walter G. Hermes, "Global Pressures and the Flexible Response," in William A. Stofft, ed., *American Military History*, Washington D.C: U.S. Army Center of Military History, 1989, pp. 591–619. For more on MC 14/3, see North Atlantic Military Committee, *Final Decision on MC 14/3: A Report by the Military Committee to the Defence Planning Committee on Overall Strategic Concept for the Defense of the North Atlantic Treaty Organization Area*, January 16, 1968.

[59] Gjert Lage Dyndal, "How the High North Became Central in NATO Strategy: Revelations from the NATO Archives," *Journal of Strategic Studies*, Vol. 34, No. 4, August 2011, pp. 560, 575–576.

[60] Dyndal, 2011, pp. 575–576.

[61] Dyndal, 2011, p. 560.

[62] "1967–1979: NATO's Readiness Increases," Supreme Headquarters Allied Powers Europe, n.d. Diego A. Ruiz Palmer, "The NATO-Warsaw Pact Competition in the 1970s and 1980s: A Revolution in Military Affairs in the Making or the End of a Strategic Age?" *Cold War History*, Vol. 14, No. 4, 2014, pp. 533–573.

Germany—particularly the Zealand group of islands and Schleswig-Holstein—and featured troops from multiple countries exercising with tracked and wheeled vehicles and helicopters.[63] Northern Wedding tested the Alliance's naval forces and concluded with a landing of forces in southern Norway.[64] In November 1978 NATO mooted an initial plan "consolidating previously separate American, British and Canadian reinforcement plans into a single SACEUR Rapid Reinforcement Plan (RRP), with an expanded focus from the Alliance's Central Region to the allies on NATO's flank regions (Denmark, Norway, Italy, Greece and Turkey)."[65]

In January 1981, "Norway and the United States [signed] an agreement for the prestocking of certain heavy equipment and for host-nation support of a U.S. marine amphibious brigade as Allied reinforcement to Norway in an emergency."[66] Around 1980 Denmark increasingly began to call for the pre-positioning of U.S. Marine equipment on its territory with the goal of obtaining "Marine support for protection against Soviet amphibious attacks on Jutland, Zealand, and the smaller Danish islands."[67] Ultimately, although Denmark accepted the Rapid Reinforcement Plan, "left-wing opposition to pre-positioning equipment for U.S. marines may explain why this was omitted from the plan."[68]

In 1982 the U.S. Department of the Navy began to formulate the Maritime Strategy, which became a formal publication in 1984. The strategy gave the U.S. Navy a more proactive role, with goals including: (1) seeking out and destroying enemy naval forces (even in high-threat areas); (2) controlling vital sea areas and protecting vital sea lines of communication; (3) denying vital areas from the enemy; (4) seizing and defending advanced naval bases; (5) blunting enemy ground attack through direct action; and (6) suppressing enemy sea commerce—all with the ultimate goal of maritime superiority.[69] Under the Maritime Strategy, the United States and the Navy adopted a much more forward-oriented posture than had previously been the case. The Maritime Strategy

> provided that the U.S. Navy and Marines would wage global coalition warfare in conjunction with the Army and Air Force and the forces of allied nations. As such, it dovetailed nicely with CONMAROPS [NATO's Concept of Maritime Operations], but in certain areas it went farther—for instance, in the taking out

---

[63] "Joint Military Exercises Begin This Week," UPI, August 29, 1982.

[64] John Jones, "A Naval Force of 35,000 Personnel and 150 Ships . . ." UPI, August 28, 1986.

[65] Ruiz Palmer, 2014, p. 555.

[66] Johan J. Holst, "Norway's Search for a Nordpolitik," *Foreign Affairs*, Fall 1981.

[67] Congressional Budget Office, *The Marine Corps in the 1980s: Prestocking Proposals, the Rapid Deployment Force, and Other Issues*, Washington, D.C.: GPO, May 1980, pp. 2–3.

[68] Olav Riste, "NATO's Northern Frontline in the 1980s" in Olav Njøstad, ed., *The Last Decade of the Cold War: From Conflict Escalation to Conflict Transformation*, Abingdon, England: Taylor and Francis, 2005, p. 307.

[69] For more information on successive iterations of the Maritime Strategy, see John B. Hattendorf and Peter M. Swartz, eds., *U.S. Naval Strategy in the 1980s: Selected Documents*, Newport, R.I.: Naval War College Press, 2008.

of Soviet SSBNs; operation of carrier battle groups (CVBGs) in coastal waters far forward, sheltered by the mountains surrounding the northern Norwegian fjords; and the concept of horizontal escalation. NATO's and the Americans' objectives in the Norwegian Sea were to repel a Soviet amphibious assault on northern Norway, support northern Norway against land threats, prevent Soviet use of facilities in Norway, and contain the Northern fleet or destroy it at sea.[70]

In 1985 the United States deployed aircraft carriers to Norwegian coastal waters for the first time since 1952 as part of the NATO exercise Ocean Safari and continued to conduct a number of high-profile exercises in the area throughout the 1980s.[71] The U.S. posture in northern Europe under the Maritime Strategy was so proactive that certain analysts condemned it as being overly aggressive and possibly leading to escalation of nuclear warfare.[72]

Through these actions the United States sought to signal its resolve to the Soviet Union. VADM Richard Allen, who commanded Ocean Safari in 1985, later stated that "by conducting exercises like Ocean Safari, the American government was sending signals to the Soviet Union, saying that the United States was on the offensive." Russian defense expert Vitaly Tsygichko further noted that the Soviet Union saw as a serious threat the Maritime Strategy, or Lehman Strategy (named after John Lehman, the Secretary of the Navy who had initiated it). At the same time, however, "the Reagan administration's comprehensive strategy, including the Maritime Strategy, made the Soviet military realize the significant technological gap that was widening between the USSR and the U.S." and gradually acknowledge that there was no way to close the gap.[73]

Lessons

When we look at the framework established in Chapter Three, U.S. efforts to deter the Soviets in NATO's Northern Flank present an interesting case of U.S. extended deterrence. (For a list of key factors in terms of the study's framework for deterrence, see Table D.1.) The variables for Category Two regarding the clarity of the defender's message existed at very high levels. The variables for Category Three regarding the credibility of the defender's threats were also present at high levels. Meanwhile, Category One, of variables regarding the Soviet Union's motivation, shows that although it considered the Northern Flank countries (and especially Norway) strategically important, it had low motivation to initiate aggression. Because of the lack of English-language sources on this topic, it is hard to gauge what individual Soviet leaders thought of the regional

---

[70] Jacob Børresen, "Alliance Naval Strategies and Norway in the Final Years of the Cold War," *Naval War College Review*, Vol. 64, No. 2, Spring 2011, p. 100.

[71] Børresen, 2011, p. 114.

[72] William M. Arkin, "Our Risky Naval Strategy Could Get Us All Killed," *Washington Post*, July 3, 1988.

[73] Kjell Inge Bjerga, *Politico-Military Assessments on the Northern Flank 1975–1990: Report from the IFS/PHP Bodø Conference of 20–21 August 2007*, p. 5.

Table D.1
Application of the Framework to the Case of the United States Deterring the USSR

Category	Variable	Level in Present Case
How motivated was the USSR?	General level of dissatisfaction with status quo and determination to create a new strategic situation.	Low. The USSR desired to transform northern Europe into a neutral, pro-Soviet area but found the general security situation tolerable.
	Degree of fear that the strategic situation was about to turn against the USSR in decisive ways.	Low. Throughout the Cold War, the Soviet Union found the general security in the region to be tolerable, if not ideal.
	Level of national interest involved in specific territory of concern.	High. The Kola Peninsula hosted the Soviet Union's Northern Fleet, which contained most of its nuclear submarines, and was also the location of Murmansk and Pechenga, the Soviets' only year-round ice-free ports. A favorable regional security situation was therefore vital to both the Soviets' deterrent and naval capabilities.
	Urgent sense of desperation or requirement to act; whether the USSR was locked into course of action.	Low. The Soviet Union maintained steady pressure on Denmark and Norway throughout the Cold War and did not feel the need to deviate from this policy.
	Degree of aggressive, reckless, risk-accepting opportunism.	Unclear. However, successive Soviet leaders appeared to consistently follow the same policy toward the region. This suggests that the personalities of individual leaders did not greatly impact policy.
	Level of motivated reasoning in play; degree of wishful thinking, misperception of basic strategic context.	Unclear. However, successive Soviet leaders appeared to consistently follow the same policy toward the region. This suggests that the personalities of individual leaders did not greatly impact policy.
Was the United States clear and explicit regarding what it sought to prevent and what actions it would take in response?	Precision in the type of aggression the United States sought to prevent.	Mixed. U.S. officials only rarely spoke about the specific types of aggressions they sought to prevent in the Northern Flank region. However, the measures the United States adopted to strengthen its presence indirectly indicated that it was concerned with ground and maritime aggression on the part of the Soviets.
	Clarity in the actions that would be taken in the event of aggression.	High. NATO exercises and U.S. posturing measures showed that it was prepared to reinforce from air and sea. Publicly known strategies like massive retaliation and flexible response also indicated other ways in which the United States was prepared to respond to Soviet aggression.
	Forceful communication of these messages to outside audiences, especially potential aggressor(s).	High. Although U.S. communication regarding Denmark and Norway specifically was rare, successive U.S. presidents were extremely clear about their commitment to NATO and to defending Europe from possible Soviet/Warsaw Pact aggression.
	Timely response to warning with clarification of interests, threats.	N/A

Table D.1—Continued

Category	Variable	Level in Present Case
Did the USSR view U.S. threats as credible and intimidating?	Actual and perceived strength of the local military capability to deny the presumed objectives of the aggression.	Mixed to high. Denmark and Norway's ability to resist potential Soviet aggression was limited because of self-imposed limitations. However, from the late 1970s onward, the United States and NATO began considerably reinforcing their ground and naval presence in the region.
	Degree of automaticity of U.S. response, including escalation to larger conflict.	High. Article V of the North Atlantic Treaty meant that, officially, an attack on Norway and/or Denmark would result in an automatic U.S. response. U.S. military strategies—including flexible response and the Maritime Strategy—allowed for horizontal escalation elsewhere in the world.
	Degree of actual and perceived credibility of political commitment to fulfill deterrent threats.	High. The Soviets saw the U.S. Maritime Strategy as a serious offensive threat.
	Degree of national interests engaged in state to be protected.	High. Loss of control of Norway would have allowed the Soviet Northern Fleet to go into open waters unhindered and would have also had negative implications for provisioning and reinforcement of NATO forces along the Central Front. The same would have been true of the loss of Denmark.
	Reputation for resolve with potential aggressor.	Mixed. U.S.-Soviet relations had gone through several phases during the Cold War, including the period of high tensions during the 1960s, détente during the early 1970s, and a return to high tensions during the 1980s. However, the two sides had not been in a situation in which the Soviet Union invaded a NATO member state.
	Degree of threat posed to attacker's values and interests by the specific responses threatened by defender.	Very high. The U.S. far-forward posture seriously threatened the Soviet Northern Fleet. Because the Northern Fleet contained a majority of the Soviets' nuclear submarines, a threat to the fleet constituted a threat to the Soviets' deterrence.

security situation, but Soviet policy remained consistent throughout the Cold War. This consistency suggests that the personalities of individual leaders had limited impact.

These aspects of the case suggest that the degree of Soviet motivation played a part in the success of U.S. extended deterrence. The Soviet Union had limited objectives for the Nordic countries that fell short of military aggression, focusing on denial of U.S. presence in the area rather than actual possession of Denmark and Norway. Over the course of the Cold War, and particularly from the late 1970s onward, the United States adopted an increasingly proactive posture in the region. The Soviet Union, on the other hand, appeared to maintain the same limited objectives of denial rather than possession. This confluence of factors appeared to contribute to the success of U.S. extended deterrence in this case.

Conclusion

U.S. deterrence efforts on NATO's Northern Flank during the Cold War show that the potential aggressor's level of motivation can contribute greatly to deterrence success or failure. The U.S. deterrence message—regarding NATO and, more specifically, Denmark and Norway—was very clear. Successive U.S. administrations were committed to the defense of Western Europe, but more concrete U.S. efforts to defend the Northern Flank in particular—such as a forward naval posture and large-scale exercises—did not begin to materialize until the late 1970s and early 1980s. In the interim, U.S. and NATO planners focused on the Central Front. During this time, the Soviet Union could have taken the opportunity to launch an attack and undermined U.S. extended deterrence. Ultimately, the Soviet Union's limited objectives of denial rather than possession, and the Soviet's resulting low level of motivation to initiate an attack against Denmark or Norway, prevented military aggression.

The question remains as to whether the United States would have achieved the same success had it strengthened its deterrent message and become more proactive in northern Europe before the 1980s. There are indications that U.S. extended deterrence might actually have come closer to failing if it undertaken this shift earlier. Northern Norway, adjacent to the Kola Peninsula and the Northern Fleet, was an area of high strategic importance to the Soviet Union; its loss could have seriously threatened the Soviet Union's naval capabilities, as well as its nuclear deterrent. Before the 1980s the United States did not seriously threaten the prevailing security situation in the area because its attention was focused elsewhere. This focus elsewhere likely contributed to the Soviet Union's limited objectives (denial rather than possession) in the region. The United States increased its forward presence during the 1980s, when the Soviets became aware of and eventually reluctantly accepted the significant and widening technological gap between the two countries' militaries. If the United States had been more proactive earlier, it might have increased pressure on the Soviets, but it also would have risked inflaming a sense of desperation; the Soviets might have come to believe they had to act before the strategic situation turned irrevocably against them in this strategically important region. The still relatively narrow capabilities gap between NATO and the Warsaw Pact militaries would have risked a prolonged and risky confrontation.

The Northern Flank case appears to suggest that, in areas where the potential aggressor has low motivation to attack, a more low-key approach—combined with active vigilance—might be sufficient to deter. A more proactive posture, on other hand, might work if the defender has an overwhelming advantage over the potential aggressor—and the aggressor is aware of this fact. Otherwise, a proactive posture might push the potential aggressor into action, causing the very act that the defender had sought to prevent in the first place.

APPENDIX E

Qualitative Case Study Analyses: Russian Aggression Against Georgia

On August 7, 2008, following months of tensions with Russia and Russian-backed separatists in the northern Georgian region of South Ossetia (see Figure E.1), the Georgian military launched artillery attacks against South Ossetia's capital, Tskhinvali. Russia responded the next day with a large-scale deployment of forces, arguing that it needed to protect its nationals in South Ossetia. Within a few days Russia had routed Georgian forces and crossed from South Ossetia into undisputed Georgian territory. By the time Russia and Georgia signed a cease-fire negotiated under the aegis of the European Union (EU) on August 12, Russia had deployed 40,000 troops in Georgia, launched a secondary offensive in Abkhazia—Georgia's other separatist region—and was closing on Georgia's capital, Tbilisi. While peace negotiations were ongoing, Moscow announced on August 26 that it officially recognized Abkhazia and South Ossetia as independent states, for the first time since the fall of the Soviet Union breaking away from the principle that former Soviet states should not be further dismembered.[1]

The weeks and months that led to the so-called Five-Day War (August 7–12, 2008) between Russia and Georgia, as well as the war itself, present two cases of deterrence involving the United States.[2] First, the United States attempted to prevent Georgia from responding militarily to provocations from South Ossetian separatists—a response that, it was thought, would automatically trigger military retaliation from Russia. Second, the United States tried to prevent Russia from crushing Georgia militarily, at a time when Russian forces had crossed into Georgia's undisputed territory and were advancing toward Tbilisi. U.S. deterrent efforts were successful in the latter case but not the former, suggesting it is sometimes easier to convince a rival (Russia) than a friend (Georgia). This case of successful deterrence, however, owes more to

---

[1] On the implications of this recognition and the parallel Russia made with the Kosovo case, see Roy Allison, "Russia Resurgent? Moscow's Campaign to 'Coerce Georgia to Peace,'" *International Affairs*, Vol. 84, No. 6, November 2009, pp. 1159–1161.

[2] While the European Union also played a role in these deterrent efforts (particularly the second one), this appendix focuses primarily on deterrence efforts by the United States, consistent with other case studies in this volume.

**Figure E.1
Map of Georgia**

Russia's lack of interest in taking over Tbilisi than to clear messaging or powerful threats on the part of the United States.

The crisis between Russia and Georgia presents some additional dynamics that resemble deterrence without fully fitting the definition. Did the United States try to deter Russia from increasing its informal control over Abkhazia and South Ossetia and supporting these region's separatist movements, in what several authors have described as "creeping annexation"?[3] While the United States saw tensions between Georgia and Russia increase over the years—specifically on the issue of Abkhazia and South Ossetia—this issue did not figure prominently on the U.S. agenda and Russia does not appear to have been warned that its involvement in Georgia's separatist provinces

---

[3] See Asmus, 2010, p. 37; and Alexander Cooley, "How the West Failed Georgia," *Current History*, October 2008, p. 343.

would trigger negative consequences. Simply put, the United States made no clear effort to deter Russia from increasing its presence in Abkhazia and South Ossetia, other than by shoring up Georgian defense forces for reasons that were largely unrelated to the breakaway provinces or Russia.[4]

Another case of potential deterrence relates to Russia's decision to intervene militarily in South Ossetia. Did the United States try to deter Moscow from resorting to force against Georgia? While the United States and Europe attempted some diplomatic efforts in spring 2008 to prevent the Russian-Georgian crisis from escalating, these efforts were overall limited.[5] The recollections of members of the administration of Present George W. Bush involved in the crisis suggest that there was a widespread expectation that Russia would seize the opportunity to crush its neighbor if given the right pretext. Russia's determination on this matter had been communicated to the United States during an encounter between Secretary of State Condoleezza Rice and Russian president Vladimir Putin in October 2006.[6] Two years later, after a steady escalation of tensions between Georgia and Russia, there was little ground for the U.S. administration to believe Moscow would have somehow softened its stance on this issue. Therefore, a Russian reaction to a Georgian attack in its disputed provinces was perceived as unlikely to be deterred, and the United States instead tried to convince Georgia to refrain from intervening militarily in Abkhazia or South Ossetia.

This appendix first reviews the origins of the August 2008 crisis between Russia and Georgia, the buildup to the war, the conflict, and its aftermath. It then proceeds to examine whether the key variables outlined in Chapter One—aggressor's motivation, clarity of defender's message, and credibility of defender's message—played a role in the success or failure of the two cases of deterrence present during this crisis. Each case concludes with an overview of additional factors (when relevant) that may have influenced the outcome of deterrence efforts, and policies to enhance deterrence that were

---

[4] U.S. military assistance to Georgia includes the Georgia Train and Equip Program, which built Georgia's forces border security and counterterrorism capabilities specifically to address the presence of radical Islamists in the Pankisi Gorge, an area of Georgia bordering Chechnya, from 2002 to 2004. It was followed by the Georgia Sustainment and Stability Operations Program, which trained Georgian units to be deployed as part of the war effort in Iraq. See U.S. Department of State, "Georgia Train and Equip Program (GTEP)," February 1, 2003; and Doug Kimsey, "Training for Iraq Boosts Security in the Caucasus," June 28, 2005, U.S. Department of Defense.

[5] Asmus, 2010, p. 145; Council of the European Union, *Report of the Independent Fact-Finding Mission on the Conflict in Georgia*: Vol. I, Brussels: Council of the European Union, September 2009, p. 33. See also Svante E. Cornell and S. Frederick Starr, "Introduction," in Svante E. Cornell and S. Frederick Starr, eds., *The Guns of August 2008: Russia's War in Georgia*, Armonk, NY: M. E. Sharpe, 2009, p. 7. Asmus, 2010, p. 154, notes that "there is little evidence that either the United States or Europe weighed in clearly and consistently at the highest levels in Moscow to warn against the consequences of Russian aggression."

[6] According to Asmus, 2010, p. 74, referring to former Secretary of State Condoleezza Rice, "[Putin] told Rice that if the Georgian leader ever moved against either Abkhazia or South Ossetia, Moscow would respond with military force and he would then officially recognize both [Abkhazia's capital city] Sukhumi and Tskhinvali."

employed successfully. Finally, we put the two cases in perspective and draw broader implications on the success of deterrent efforts in general.

Background

The 2008 war represented the culmination of tensions that had built up between Georgia and Russia over the years. In the wake of the conflicts that pitted separatists in South Ossetia (1991–1992) and Abkhazia (1992–1994) against Georgian forces, Russia established a presence in these regions through the deployment of peacekeepers.[7] Regaining full control of these two provinces, which represented one-fifth of Georgia's entire territory, was high on the Georgian leadership's agenda. Shortly after Vladimir Putin became president, Russia started putting more pressure on Georgia while deepening its ties with the separatist provinces.[8] Moscow supported South Ossetian political hardliners who opposed accommodation with Tbilisi, and after 2002 Moscow accelerated its "passportization" policy consisting in granting Russian citizenship to an increasingly large number of inhabitants of these regions.[9]

Moscow perceived the 2004 Rose Revolution that saw pro-EU and pro-NATO politician Mikheil Saakashvili remove Georgian president Edvard Shevarnadze from power as a threat both to its internal stability and to its influence over its most strategic neighbors. Georgia, like Ukraine, applied for a Membership Action Plan (MAP) that would provide a clear path for NATO membership. Yet the April 2008 NATO Summit in Bucharest, where these MAPs were discussed, showed divisions between NATO members. While the Bush administration supported Georgia and Ukraine's request, German chancellor Angela Merkel and French president Nicolas Sarkozy were more reluctant, partly out of concern that Russia would perceive such a move as a provocation.[10] Eventually the allies settled on a compromise, whereby no MAP would be granted at that point, but NATO publicly announced that these two countries would eventually become members, without further specifications. This equivocal outcome

---

[7] Russia contributed to a Georgian-Russian-Ossetian joint force in South Ossetia, and to a Commonwealth of Independent States (CIS) force in Abkhazia. In that latter case, however, Russia was the only CIS country to deploy peacekeepers (Council of the European Union, 2009, para. 6).

[8] Andrei Illarionov, "The Russian Leadership Preparation for War," in Svante E. Cornell and S. Frederick Starr, eds., *The Guns of August 2008: Russia's War in Georgia*, Armonk, N.Y.: M. E. Sharpe, 2009, p. 51.

[9] Stent, 2014, p. 169; Council of European Union, 2009, p. 18. For a more detailed analysis of this issue, see Kristopher Natoli, "Weaponizing Nationality: An Analysis of Russia's Passport Policy in Georgia," *Boston University International Law Journal*, Vol. 28, Summer 2010, pp. 389–417.

[10] For an analysis of the discussions on a MAP for Georgia at the Bucharest Summit, see, for example, Asmus, 2010, pp. 111–140; Paul Gallis, *The NATO Summit at Bucharest, 2008*, Washington, D.C.: Congressional Research Service, May 5, 2008, pp. 5–6; and Travis L. Bound and Ryan C. Hendrickson, "Georgian Membership in NATO: Policy Implications of the Bucharest Summit," *Journal of Slavic Military Studies*, Vol. 22, No. 1, 2009, pp. 20–30.

disappointed Georgia without assuaging Russian concerns.[11] Putin had issued clear warnings to NATO against further expansion, presenting it as a "provocation" directed against Russia at the February 2007 Munich Security Conference—an accusation he reiterated at the April 2008 Bucharest Summit.[12] Another source of tensions between Georgia and Russia was Tbilisi's efforts to be part of a new energy corridor that would connect Azeri resources (and, in the long term, Central Asian resources) to European markets.[13]

Russia and Georgia experienced recurring crises. In 2006, the expulsion by Georgia of four Russian spies prompted a series of retaliatory measures from Russia, including the boycott of Georgian wine and mineral water, as well as the expulsion of Georgian migrant workers.[14] Incidents taking place between Russia and Georgia usually centered on the frozen conflicts of Abkhazia and South Ossetia. In one instance in August 2007, a Russian jet fired a missile at a Georgian radar station near Gori, about 20 miles from Tskhinvali.[15] In April 2008, shortly before leaving the Russian presidency to Dmitry Medvedev, Vladimir Putin signed a presidential decree establishing official relations with Abkhaz and South Ossetian self-declared authorities, which Georgia did not recognize. That same month, a Russian MiG-29 shot down a Georgian unmanned aerial vehicle that was conducting a surveillance mission over Abkhazia. At that point, Abkhazia was seen as the most likely flashpoint between Georgia and Russia. Over the summer, however, the focus of tensions shifted to South Ossetia, with the confrontation between Georgian forces and South Ossetian separatists intensifying.[16] Russia and Georgia accused each other of preparing for war, with Georgia suspecting Russia of building up South Ossetian militias with "volunteers" coming from North Ossetia, as well as providing them with military equipment through the Roki Tunnel that connects the two regions.[17] Georgia and Russia each held a series of military exercises. On Georgia's side, the exercise was led by the United States and included troops from Armenia, Azerbaijan, and Ukraine in addition to Georgia. Russia's exercise took place on the other side of the Georgian border.[18]

During the night of August 7–8, Georgia launched an artillery attack against Tskhinvali. Russia claimed that the attack killed two Russian peacekeepers, and

---

[11] Stent, 2014, p. 167.

[12] Vladimir Putin, quoted in Fiona Hill and Clifford G. Gaddy, *Mr. Putin: Operative in the Kremlin*, Washington, D.C.: Brookings Institution Press, 2013, p. 307.

[13] Asmus, 2010, pp. 9, 57.

[14] Stent, 2014, p. 169.

[15] Misha Dzhindzhikhashvili, "Georgia: Russian Jet Fired Missile," *Washington Post*, August 8, 2007.

[16] Stent, 2014, p. 170; Council of the European Union, 2009, pp. 18–19.

[17] Allison, 2009, p. 1147.

[18] Stent, 2014, pp. 169–170.

launched its own offensive on the morning of August 8. Russia's forces quickly overran the positions of the Georgian forces and crossed from South Ossetia into the rest of Georgia. On August 12, while Russian forces were closing in on Tbilisi, Russian president Dmitry Medvedev agreed to a cease-fire negotiated by French president Sarkozy.[19] By that point, the confrontation led to dramatic results for Georgia. After opening a second front in Abkhazia, Russia controlled the city of Zugdidi in western Georgia, the Zenaki military base, and the Port of Poti on the Black Sea, effectively cutting the country in two. Ethnic Georgian villages in South Ossetia were systematically looted and burned, their residents sent fleeing.[20] The Abkhaz self-proclaimed government forces used the opportunity to take the Upper Kodori Valley, which had been controlled by Georgia.[21] On August 26, Russia formally recognized Abkhazia and South Ossetia as sovereign territories. Even after Moscow pulled its forces out of Georgia's undisputed territory, it maintained a larger-than-before presence in both Abkhazia and South Ossetia, and large numbers of ethnic Georgians displaced during the war were never able to return to South Ossetia.

U.S. Attempts to Deter Georgia from Using Force in South Ossetia

As tensions built up between Russia and Georgia, first around Abkhazia, then South Ossetia, the United States tried to deter its Georgian ally from intervening in the breakaway provinces—a move the Bush administration believed would provide Russia with a pretext to respond forcefully. In her memoir, then–Secretary of State Condoleezza Rice recalls about Saakashvili that "we all worried that he might allow Moscow to provoke him to use force."[22] U.S. efforts to prevent Saakashvili from confronting militarily Russia in South Ossetia represent a case of failed deterrence. On August 7, Georgian forces shelled Tskhinvali, prompting Russia to claim that its nationals—whether Russian peacekeepers or South Ossetians with Russian citizenship—required protection, thus causing the military invasion of South Ossetia before advancing into Georgia.

The Aggressor's Motivation

For years, Saakashvili had wanted to bring both Abkhazia and South Ossetia back under Tbilisi's authority, and reportedly considered plans for military offensives in both regions.[23] U.S. officials were aware of that fact and, according to one account, communicated to Saakashvili that the good relations enjoyed by the United States and

---

[19] France was, at the time, holding the presidency of the European Union.
[20] Human Rights Watch, "Georgian Villages in South Ossetia Burnt, Looted," August 12, 2008.
[21] Allison, 2009, p. 1158.
[22] Rice, 2011, p. 685.
[23] Asmus, 2010, pp. 79–81.

Georgia did not mean that the former would support military action against separatists in Abkhazia and South Ossetia.[24] Angela Stent wonders whether the April 2008 NATO Summit, where Georgia ended up not getting the MAP it had been hoping for, may have put pressure on Saakashvili to resolve the frozen conflicts that held back Georgia's potential future membership.[25] Yet according to Ronald Asmus, Saakashvili had dropped all plans for a potential offensive in South Ossetia as early as 2006, nonetheless keeping the military option open for Abkhazia, which was perceived to be of greater strategic importance—and where, in spring 2008, a conflict with Russia seemed imminent.[26]

Saakashvili's personality—variously described as at times "impulsive,"[27] "combustible,"[28] and "brash and hyperkinetic"[29]—is commonly mentioned as playing a role in setting Georgia on a path to war. As one journalistic account mentions, "By 2008, Saakashvili's hubris led him to challenge the Russians openly, boasting of how 'my tanks' and 'my radars' were fully capable of confronting the Bear."[30] Another mentions scores of diplomats blaming the war on Saakashvili's "hubris."[31] In her memoir, Rice acknowledges concerns from the U.S. administration that Saakashvili may have believed that he would be able to replicate the success he had in Ajara, a province in the southwest of Georgia that did not recognize the authority of Tbilisi until, in spring 2004, a combination of sanctions and support for local protests brought it back within Georgia.[32]

One prevailing explanation for the war holds that Saakashvili fell in a "trap" set by Russia.[33] According to this theory, Russia's activities in Abkhazia and South Ossetia were aimed at luring Georgia into attacking South Ossetia, thereby providing Russia

---

[24] Cooper, Chivers, and Levy, 2008.

[25] Stent, 2014, p. 168.

[26] Asmus, 2010, pp. 80–85.

[27] Rice, 2011, p. 685.

[28] Asmus, 2010, p. 13.

[29] Cooper, Chivers, and Levy, 2008.

[30] Andrew Cockburn, "The Bloom Comes Off the Georgian Rose," *Harper's*, October 31, 2013.

[31] Wendell Steavenson, "Marching Through Georgia: Has Mikheil Saakashvili Overreached?" *New Yorker*, December 15, 2008.

[32] Rice, 2011, p. 685. On the Ajara crisis, and the limits of the comparison between Ajara, Abkhazia, and Ossetia, see International Crisis Group, *Saakashvili's Ajara Success: Repeatable Elsewhere in Georgia?* Brussels: International Crisis Group, August 18, 2004.

[33] Asmus, 2010, p. 48; Cornell, 2008, p. 312, notes that "the only thing Saakashvili might be blamed for is falling into a trap that Russia had prepared for months"; Robert Kagan, "Putin Makes His Move," *Washington Post*, August 11, 2008, points to Russia's responsibility for "encouraging South Ossetian rebels to raise the pressure on Tbilisi and make demands that no Georgian leader could accept. If Saakashvili had not fallen into Putin's trap this time, something else would have eventually sparked the conflict." Cornell and Starr, 2009, p. 9, similarly argue that Russia had planned for precisely that war for months if not years and if Georgia had not stumbled onto that trap, Russia would have found another way to justify its intervention.

with a pretext to attack its Western-leaning, independent-minded neighbor. In the months that preceded the war, the United States warned Saakashvili to avoid falling into that trap, and tried to secure from him a commitment that he would not use military force in the separatist provinces—to no avail.[34]

This explanation supposes that Saakashvili purposely chose military confrontation rather than diplomacy in a repertoire of policy options. However, the Georgian leader argues instead that he had no option other than confronting Russia, with whom a war was inevitable.[35] A number of accounts provide some support to the Georgian argument. Pavel Felgenhauer notes that following Georgia's rapprochement with NATO, "military action in support of separatists in Abkhazia and South Ossetia is being seriously contemplated" in Moscow.[36] Russian defense chief General Yury Baluyevskiy had communicated to his NATO counterparts that a war in Georgia over the summer was likely unless some steps were taken to de-escalate the situation.[37] Finally, the speed and quality of the Russian counteroffensive against Georgian troops suggests that Moscow had been more than contemplating a possible war[38]—it had actively prepared for it, as illustrated by the Caucasus 2008 military exercise that took place in July 2008 and played out a scenario similar to the actual operations that Russian forces would undertake in August.[39] Based on these elements, it is likely that the Georgian leadership was bracing itself for a confrontation that it saw as largely inevitable.[40]

Additionally, on August 7, Georgia claimed that the war was not just inevitable but already underway.[41] More specifically, it argued that artillery attacks were only

---

[34] For a critique of the "Russian trap" hypothesis, see Samuel Charap and Timothy J. Colton, *Everyone Loses: The Ukraine Crisis and the Ruinous Contest for Post-Soviet Eurasia*, Adelphi Series, Vol. 56, London: International Institute for Strategic Studies, 2016, pp. 92–93. Allison, 2009, p. 1161, similarly agrees that Russia's attack against Georgia "is unlikely to have been the expression of some master plan, or even the product of calculated broad consultation, even if military planning of the operation itself took place in advance."

[35] Asmus, 2010, p. 144, describes Saakashvili as "convinced that Putin was preparing for war" after their meeting in Moscow in February 2008.

[36] Pavel Felgenhauer, "Moscow Ready for Major Confrontations with Pro-Western Georgia and Ukraine," Jamestown Foundation, *Eurasia Daily Monitor*, Vol. 5, No. 117, June 19, 2008.

[37] Asmus, 2010, p. 149.

[38] For arguments that Russia's offensive against Georgia was well-planned in advance, see Cornell, 2008, p. 311; Allison, 2009, p. 1149; and Asmus, 2010, p. 166. Charap and Colton, 2016, pp. 92–93, and Asmus, 2010, p. 168, argue, however, that the timing of the Georgian offensive probably came as a surprise for Russia.

[39] Carolina Vendil Pallin and Fredrik Westerlund, "Russia's War in Georgia: Lessons and Consequences," *Small Wars and Insurgencies*, Vol. 20, No. 2, June 2009, p. 400.

[40] While this may have been the Georgian side's perception—likely heightened by the months of tensions with Russia that preceded—this does not necessarily mean that Russia was ready to attack when it did. Charap and Colton, 2016, 93, argue that "on 7 August, Russia had only minimal forces present near the Roki Tunnel leading from Russia to South Ossetia. Medvedev was on a cruise on the Volga; Putin, now prime minister, was at the Beijing Olympic Games; and the defence minister was on vacation on the Black Sea coast. In short, both sides were preparing for a war, but neither was planning for the war that actually happened."

[41] Steavenson, 2008.

launched on the night of August 7–8 after Russia started moving heavy military equipment toward South Ossetia through the Roki Tunnel.[42] To support its narrative, the Georgian side supplied phone intercepts purportedly showing that some heavy Russian equipment had already crossed the tunnel by the evening of August 7. Russia, however, argued that it was simply a routine rotation of its peacekeeping force.[43]

If Saakashvili did believe that war with Russia was highly likely or even already underway, his options besides a military attack were indeed limited. Abkhazia and South Ossetia figured prominently in the 2004 campaign that had gotten him elected, and on his agenda; he was unlikely to survive politically if Russia took over South Ossetia.[44] Even without taking into account the political salience of South Ossetia, not responding in kind to a Russian attack would have inflicted damage beyond repair to Saakashvili's credibility at a time when his crackdown on the peaceful protests of November 2007 and subsequent state of emergency in Tbilisi eroded his popularity. Finally, if the Russian attack did not stop in South Ossetia but continued instead toward Tbilisi (as it eventually did), it would almost certainly remove Saakashvili from power.[45] As Vano Merabishvili, then Georgia's minister of the interior, put it in an interview after the fact,

> We were faced with a situation where there was no choice. Or are you saying I had to stand by in Tbilisi and wait for the Russian tanks? Maybe we gained some time by acting fast. What would have happened if the Russian tanks invaded, and without resistance got to Tbilisi?[46]

Georgia's motivation to launch its attack in South Ossetia was therefore extremely strong, since all other options led to riskier or worse outcomes for Saakashvili. As Asmus puts it, "The Georgian decision to use force was made at the last second by a leader who felt cornered."[47]

The Clarity of the Defender's Message

In their memoirs, U.S. officials who communicated with Saakashvili as tensions with Russia mounted emphasize their efforts to convey to the Georgian leader that he should

---

[42] Allison, 2009, p. 1148. Steavenson, 2008, reports a story from the Georgian Minister for reintegration, Temuri Yakobashvili, that supports this version: "Later that evening [August 7], Yakobashvili said, Saakashvili received a call that seemed to confirm reports of Russian units moving south through the Roki Tunnel, the only road between Russia and South Ossetia. He described Saakashvili putting down the receiver: 'He got pale. I asked him what had happened, and he said, "They're moving."'"

[43] C. J. Chivers, "Georgia Offers Fresh Evidence on War's Start," *New York Times*, September 15, 2008.

[44] Cooper, Chivers, and Levy, 2008.

[45] Asmus, 2010, p. 10; Cooper, Chivers, and Levy, 2008.

[46] Merabishvili interview, 2010. Cornell, 2008, p. 312, argues that Georgia's use of force did slow down the progression of the Russian advance, possibly giving more time for negotiations to succeed before the Russian forces could reach Tbilisi.

[47] Asmus, 2010, p. 49.

not use force in the breakaway regions, and that if he did it, he should not expect military support from the United States. Condoleezza Rice, for instance, recalls,

> Finally I thought I'd better get tougher. "Mr. President, whatever you do, don't let the Russians provoke you. You remember when President Bush said that Moscow would try to get you to do something stupid. And don't engage Russian military forces. No one will come to your aid, and you will lose," I said sternly. He got the point, looking as if he'd just lost his last friend. I tried to soften what I'd said by repeating our pledge to defend Georgia's territorial integrity—with words. He asked if I'd say so publicly. I did, avoiding any language that might be misinterpreted as committing us to Georgia's defense with arms.[48]

This recollection, if accurate, suggests Saakashvili understood clearly that the United States would not come to his help in case of a confrontation with Russia. Then–National Security Adviser Stephen J. Hadley similarly remembers, "We made all kinds of signals to Putin to stay out and the president made all kinds of signals to President Saakashvili not to provoke Putin. I remember he said, 'Don't provoke Putin. You can't handle him and we will not be able to save you from Putin.'"[49] According to one journalistic account, President Bush took aside Saakashvili during the NATO Summit in Bucharest to tell him, "The U.S. would not start World War Three on his behalf."[50] Stent, based on interviews with the White House Chief of Staff and the U.S. ambassador to Georgia at the time, notes that Bush had "explicitly warned Saakashvili not to let the Russians provoke him and not to use force to take back the regions, making it clear that the United States would not come to Georgia's rescue if it did."[51] The message that Georgia should not go into South Ossetia was repeated by then–Assistant Secretary of State for European and Eurasian Affairs Daniel Fried to Georgia's foreign minister Eka Tkeshelashvili as late as August 6.[52]

A few accounts dispute the clarity of the U.S. message to the Georgian leadership. According to journalist Andrew Cockburn, who quotes a former U.S. national security official, Vice President Dick Cheney may have had a different message from the rest of the administration: "'At best Georgia would win, in which case Russia would fall apart,' the official told me, 'and at worst the spectacle of Russia crushing little Georgia would reinforce Russia's reputation as the cruel Goliath. So Cheney was telling Misha, 'We have your back.'"[53] There has also been some speculation that Secretary Rice's public statement while visiting Tbilisi in July 2008, "We always fight for our friends,"

---

[48] Rice, 2011, p. 686.

[49] Labott, 2014.

[50] Cockburn, 2015.

[51] Stent, 2014, p. 168.

[52] Cooper, Chivers, and Levy, 2008.

[53] Cockburn, 2013.

may have been misread by Georgia.⁵⁴ While clearly made in reference to an upcoming NATO meeting in December and the issue of membership, this statement may have been construed, given the context of escalating tensions, as indicating a more forceful type of support.⁵⁵ Yet Asmus's well-documented account of the war concludes,

> Speculation over whether Washington had given Tbilisi some kind of green light misses the point. No senior Georgian official has actually ever suggested that Washington did so. On the contrary, they all admit that warnings had been given repeatedly by senior American and European officials.⁵⁶

One exception—and apparently an isolated case—is that of Erosi Kitsmarishvili, then the Georgian ambassador to Moscow. During a hearing held by the Georgian Parliament in November 2008, he mentioned a "green light" given by the Bush administration to Saakashvili to take military action against South Ossetia.⁵⁷

The Credibility of the Defender's Threats

The question of whether Saakashvili believed that the United States would hold to its promise not to intervene in the case of a war between Georgia and Russia may seem of marginal importance if, as argued above, the Georgian leader believed the Russian intervention was underway and he had no choice but to respond to Russia. Yet answering that question would still shed some light on how risky a course of action Saakashvili thought this would be.

The degree of clarity of the U.S. deterrent message has direct implications for the credibility of the threat. If Saakashvili believed that members of the U.S. administration held different views on whether to support a Georgian incursion into South Ossetia, then he may have believed that these individuals—Vice President Cheney, mainly—would take a part in the decision to come to Georgia's help if its situation became too desperate, and that their views could have prevailed.

Georgian leadership may have found the U.S. message of nonassistance difficult to believe because it was inconsistent with the very positive bilateral relationship that the two countries enjoyed at the time, including in the military domain. With a population of less than four million, Georgia provided the third largest foreign contingent (after the United States and the United Kingdom) to the war effort in Iraq. In 2005,

---

⁵⁴ Condoleezza Rice, "Remarks with Georgian President Mikheil Saakashvili," July 10, 2008.

⁵⁵ This is, for instance, the interpretation of Hill and Gaddy, 2013, pp. 387–388, who note, "In Georgia in 2008 . . . Putin called the West's bluff about standing by its friends—which is what U.S. Secretary of State Condoleezza Rice told Georgian president Mikheil Saakashvili the West would to during a visit to Tbilisi shortly before the August war."

⁵⁶ Asmus, 2010, pp. 30–31.

⁵⁷ Mike Bowker, "The War in Georgia and the Western Response," *Central Asian Survey*, Vol. 30, No. 2, June 2011, p. 198; Olesya Vartanyan and Ellen Barry, "Ex-Diplomat Says Georgia Started War Against Russia," *New York Times*, November 25, 2008.

100,000 Georgians rallied for President Bush's speech on Tbilisi's Freedom Square.[58] Georgia received from the United States over $700 million in direct governmental assistance from 2003 to 2006 and close to another $300 million from the Millennium Challenge Corporation program.[59] As of 2008, 100 U.S. military advisers were present in Georgia, working with Georgian armed forces.[60] The Georgian leadership may have believed—or at least hoped—that if their country was on the verge of being crushed by the Russian military, the United States would react and provide some form of support.

The timing of the crisis in relation to the U.S. electoral cycle may also have weakened the credibility of U.S. messaging. As tensions mounted in 2008 between Russia and Georgia, the administration of President George W. Bush was coming to an end; all eyes were on the political campaign and the upcoming election and, with regard to foreign policy, on Iraq and the Middle East.[61] The upcoming change of administration may have signaled a shorter attention span on the part of U.S. officials involved in the bilateral relationship with Georgia, and with managing Russo-Georgian tensions. The discussion on whether or not to grant MAP status to Ukraine and Georgia, for instance, played a key role in the rising tensions between Russia and Georgia, yet as Rice notes in her memoirs, "I'd assumed that we would not push this step within the Alliance before the President left office."[62] Presidential candidate John McCain's political views were highly critical of Russia, and he and his team had struck a friendship with Saakashvili.[63] Yet while the prospect of a potential President McCain might have made the U.S. position of nonintervention less compelling for Georgia, this would not explain why Georgia chose to strike Tskhinvali in the summer rather than after the hypothetical election of McCain, when U.S. support for Georgia would have likely been even stronger.[64]

Lessons

This case suggests that the framework outlined in Chapter Three is applicable to instances of countries deterring allies, rather than just rivals or enemies, from pursuing a certain course of action. Most of the variables highlighted in Chapter Three as relevant for deterrence were present at a high level in this case (see Table E.1[65]). Georgia's

---

[58] Cooper, Chivers, and Levy, 2008.

[59] International Crisis Group, *Georgia: Sliding Towards Authoritarianism?* Europe Report No. 189, December 19, 2007, p. 13.

[60] Cooper, Chivers, and Levy, 2008; Stent, 2014, p. 169.

[61] Asmus, 2010, pp. 2–3; Cornell and Starr, 2009, p. 7.

[62] Rice, 2011, p. 670.

[63] See, for example, Michael Cooper and Elisabeth Bumiller, "War Puts Focus on McCain's Hard Line on Russia," *New York Times*, August 11, 2008. Asmus, 2010, p. 58, notes, "In January 2005, Senators Hillary Clinton and John McCain jointly nominated [Saakashvili] for the Nobel Peace Prize."

[64] Incidentally, this may have emboldened Russia to intervene in Abkhazia and South Ossetia before the election.

[65] The qualitative measures provided in Table E.1 (very low, low, mixed, high, very high) are only notional.

Table E.1
Application of the Framework to the Case of the United States Deterring Georgia

Category	Variable	Level in Present Case
How motivated was Georgia?	General level of dissatisfaction with status quo and determination to create a new strategic situation.	Low. Saakashvili may have had the ambition to retake South Ossetia eventually, but it is unlikely he would have done it at that specific time.
	Degree of fear that the strategic situation is about to turn against them in decisive ways.	Very high. Strong indications that Georgia saw a Russian attack as imminent or underway.
	Level of national interest involved in specific territory of concern.	Very high. Abkhazia and South Ossetia were important for Saakashvili politically. If a Russian invasion was indeed underway, a quick strike could allow Georgia to gain time and limit losses.
	Urgent sense of desperation or requirement to act; whether aggressor is locked into course of action.	Very high. Strong indications that Georgia saw itself with no other option than to at least try to limit the blow of the Russian offensive.
	Degree of aggressive, reckless, risk-accepting opportunism.	High, based on descriptions of Saakashvili's personality.
	Level of motivated reasoning in play; degree of wishful thinking, misperception of basic strategic context.	Unclear. Saakashvili may have believed that the United States would still intervene in some way if it looked like Russia was about to crush Georgia.
Was the United States clear and explicit regarding what it sought to prevent and what actions it would take in response?	Precision in the type of aggression the United States sought to prevent.	Very high. U.S. officials explicitly warned Saakashvili not to take any offensive measures against Abkhazia and South Ossetia.
	Clarity in the actions that would be taken in the event of aggression.	Mixed. U.S. officials made clear that the United States would not come to Georgia's rescue if it attacked first. However, Georgia may have believed the United States would intervene if Russia was the attacker and Georgia was only responding to an attack already underway.
	Forceful communication of these messages to outside audiences, especially potential aggressor(s).	Very high. The U.S. warning not to attack was delivered by high-level officials directly to Saakashvili, on several occasions, in different forums.
	Timely response to warning with clarification of interests, threats.	Very high. The United States reiterated its warning to the Georgian leadership on several occasions, including shortly before the beginning of the war, and were specific about what action Georgia should not take.
Did Georgia view U.S. threats as credible and intimidating?	Actual and perceived strength of the local military capability to deny the presumed objectives of the aggression.	N/A. The United States tried to deter Georgia from intervening not by threatening use of force but by threatening not to use force for its defense.
	Degree of automaticity of U.S. response, including escalation to larger conflict.	Low. There was nothing "automatic" in the U.S. lack of support to Georgia. Instead, U.S. officials had on many occasions mentioned the U.S.-Georgia friendship and close relationship.

Table E.1—Continued

Category	Variable	Level in Present Case
	Degree of actual and perceived credibility of political commitment to fulfill deterrent threats.	Mixed. Not intervening in Georgia was the majority view in the U.S. administration, but it was not unanimous. It is unclear whether this may have led the Georgian leadership to believe that the United States might change its mind, especially if Russia was successfully portrayed as the aggressor.
	Degree of national interests engaged in state to be protected.	N/A. In this case, the state to be protected was also—paradoxically—the aggressor. The United States did not want Georgia to get into a war because it knew Georgia could not win.
	Reputation for resolve with potential aggressor.	High. There are indications that U.S. (and European) pressures had prevented Saakashvili from intervening in Abkhazia in April 2008.
	Degree of threat posed to attacker's values and interests by the specific responses threatened by defender.	Very high. Georgia could not expect to win a war against Russia. Its only hope in such a confrontation was that Russia would pull back out of concern that the United States would support Georgia militarily; hence Saakashvili's repeated calls to U.S. officials for help after the war started.

motivation to attack Tskhinvali was very high, and the U.S. message was very clear and reasonably credible. Deterrence failed because the first factor—Georgia's motivation—was so strong that the Georgian leadership accepted all potential costs. The variable that plays a fundamental role in this dynamic is Tbilisi's belief that it was locked into a course of action. While other variables pertaining to the aggressor's motivation can be balanced against the costs of aggression, costs become largely irrelevant if the aggressor believes that there is no other option besides aggression.

Since Georgia's motivation was so high, it is difficult to assess how the United States could have improved its messaging, since it was most likely to fail in any case. Two elements, however, stand out. First, it seems that the U.S. message was at odds with the very positive bilateral relationship that the United States and Georgia enjoyed, to the point almost of cognitive dissonance. Saakashvili had to reconcile the fact that Bush had supported his country, both politically and financially, and would nonetheless let Russia potentially invade and occupy it, as well as install someone more favorable to Moscow's views at its helm. There might have been some wishful thinking on the part of Saakashvili that the United States would have intervened, somehow, to prevent such an outcome if it seemed imminent. The United States could have made its message to Georgia clearer by, for instance, pulling out temporarily some of or all its military advisers, claiming that in the current state of tensions Washington feared for their safety. Such a move would have communicated to Georgia a commitment to not being involved in any potential confrontation between Russia and Georgia, and may have put more pressure on Georgia to sign the pledge of nonuse of force that Washington tried to secure from its ally. Second, the U.S. deterrent message may have been weaker because of its timing in relation to the U.S. electoral cycle. The outgoing administration may have been less interested in launching some ambitious diplomatic initiatives to stop the escalation of tensions between Moscow and Tbilisi in 2008 than it would have been at an earlier time—yet, this is the one step (arguably, a demanding one) that could have potentially prevented the perception of a "lock-in" that caused Georgia to launch its attack.

U.S. Attempts to Deter Russia from Continuing Its Advance on Tbilisi

U.S. attempts at stopping Russia's advance toward Tbilisi after its military forces routed Georgia's represent a second case of deterrence. At first glance, this case seems successful. The United States indicated to Russia that pursuing its offensive would be costly, and Russia stopped. Yet successful deterrence would have required Russia to have had the intention to take Tbilisi, and to have modified its plans specifically in reaction to the U.S. message in order to avoid the costs the United States threatened to impose. A closer look suggests instead that Russia's motivation to reach Tbilisi and possibly remove Saakashvili from power was low, and that it did not see the U.S. threat as credible—although evidence on this second point is more limited. U.S. efforts at stopping Russia's advance are therefore not a case of successful deterrence. It is not failed deterrence either; that would have entailed Russia *not* stopping its advance in response to the U.S. message. Rather, it is a case where the defender may have believed it was deterring an aggressor effectively while, in reality, there was little to deter.

The Aggressor's Motivation

How intent was Russia on going to Tbilisi and possibly removing Saakashvili from power? In his memoir, Dick Cheney indicates that he believed at the time that Russia had a plan for taking over Tbilisi, noting about the Russian forces headed toward the capital,

> They moved with the kind of speed that strongly suggested significant advance planning, leaving the impression that Putin had been planning such an attack for some time. That, together with the fact that the Russian response was far in excess of what was required if the goal was simply to protect Russians in South Ossetia and Abkhazia, suggested to me and many others that Putin was intent on reasserting Russian influence in its former republics.[66]

---

[66] Dick Cheney with Liz Cheney, *In My Time: A Personal and Political Memoir*, New York: Threshold Editions, 2011, p. 513.

According to Condoleezza Rice's memoir, on August 11 Russian foreign minister Sergey Lavrov told her that Russia wanted Saakashvili removed from power.[67] Accounts abound of the intense animosity between Vladimir Putin and the Georgian president,[68] fueled by a fundamental disagreement on the geostrategic orientation that Georgia should take—pro-European and NATO or closer to the Russian orbit.

Yet even if Russia aimed to pursue regime change in Georgia, it does not mean that it was willing to fulfill it at any cost or even that it was a priority. During the five days that the war lasted, Russia achieved a number of other important objectives. It inflicted extensive damage to Georgian military capabilities; it reinforced its presence in Abkhazia and South Ossetia; it reminded NATO members that a Georgian membership could be a liability as long as the conflicts in Abkhazia and South Ossetia were not solved;[69] and it made clear that it could invade Georgia if it chose to do so without much of a response from Tbilisi's allies—a powerful message that resonated not just in Georgia but also in other former Soviet countries looking west.[70]

The Russian side argues that taking Tbilisi was never an objective. Lieutenant General Anatoly Khrulev, who was commanding the Russian offensive, stated in an interview that he had never been ordered to take Tbilisi. Rather, the chief of the general staff of the armed forces of Russia, General Nikolai Makarov, told him to stop his troops' advance about 45 miles from the capital city.[71] In another interview, then–Russian president Dmitry Medvedev, when asked why Russia did not "go to Tbilisi," provided the following response:

> You know, I think that the objective of the peace enforcement operation, which lasted five days, was achieved. The purpose of this operation was not to capture Tbilisi or any other city. We just needed to stop the aggression unleashed by Saakashvili. Moreover, I am not the judge and not the executioner, again, the judgment of Saakashvili and his fate should be determined by the people with a vote or other means, as it sometimes happens in history. But my plan at the time was not—and I can honestly say, I think I did the right thing—to overthrow Saakashvili by force, although it would have been very easy.[72]

In that same interview, Medvedev justifies the Russian response—which had generally been characterized by the international community as disproportionate to the

---

[67] Rice, 2011, p. 688.

[68] See, for example, Cornell, 2008, p. 312; Steavenson, 2008; and Asmus, 2010, pp. 70–71.

[69] Rice, 2011, p. 691.

[70] Hill and Gaddy, 2013, p. 331.

[71] Lieutenant General Anatoly Khrulev, interview, April 8, 2014, translated from Russian by Samuel Charap.

[72] Dmitry Medvedev, interview, August 5, 2011, translated from Russian by Samuel Charap.

Georgian attack—as necessary to "to dismantle, to destroy the military machine of Georgia, so it couldn't strike at the civilians, the residents of Abkhazia, Ossetia and the Russian Federation."[73] With these aims achieved in five days, Russia did not need to push its advantage.

Russia's acquiescence to the diplomatic process may have also been prompted by a realization that a longer war might have been too costly. The Georgia intervention revealed major shortcomings in Russian forces and equipment and provided much of the impetus for the ambitious reform of the Russian military that begun that same year under the leadership of defense minister Anatoliy Serdyukov.[74] Russia also found itself quickly isolated internationally, in spite of its attempts to justify its intervention on the grounds that it was protecting the South Ossetian population from genocide.[75] This isolation had not just a reputational cost but also an economic one, with a decrease in foreign investors in Russia.[76] With the humiliation inflicted to Georgian forces and the damage incurred, Moscow could hope that Saakashvili would not survive politically—in an interview on August 31, 2008, Medvedev called Saakashvili a "political corpse"[77]—leaving it to the Georgian people to rid them of their leader. Pursuing a diplomatic solution gave Russia much of what it wanted without the costs of regime change.

The Clarity of the Defender's Message

As reports of Russian troops crossing into Georgia and possibly heading for Tbilisi started reaching U.S. officials, so did calls from Saakashvili claiming that Putin wanted to remove him from power and asking for U.S. support. The United States took a number of measures that Rice described as "visible help" for "our friends" and that included the sending of a destroyer in the Black Sea; delivery of humanitarian aid by military planes; the return of the Georgian unit deployed in Iraq; and U.S. involvement in the international negotiations to defuse the crisis, with Rice traveling to France to meet with the lead negotiator, French president Nicolas Sarkozy.[78]

---

[73] Medvedev interview, 2011.

[74] On this issue see, among others, Roger N. McDermott, "Russia's Conventional Armed Forces and the Georgian War," *Parameters*, Spring 2009, pp. 67–73; Pallin and Westerlund, 2009, pp. 404–413; and Ariel Cohen and Robert F. Hamilton, *The Russian Military and the Georgian War: Lessons and Implications*, Carlisle, Penn.: Strategic Studies Institute, U.S. Army War College, June 2011.

[75] For a detailed argument on Russia's justifications for intervention, see Allison, 2009, pp. 1151–1155.

[76] McDermott, p. 67; Asmus, 2010, p. 220.

[77] Dmitry Medvedev, quoted in Ellen Barry, "Russia President Dismisses Georgia's Leader as a 'Political Corpse,'" *New York Times*, September 2, 2008. Saakashvili's approval rating, however, increased in the wake of the August war; see Steavenson, 2008.

[78] Rice, 2011, p. 689.

Hadley credits these measures, which he describes as "signaling," for preventing Russia from a full-blown attack against the Georgian regime: "The road was open. We kept them from going to Tbilisi and overturning the democratic government."[79] It is not clear, however, what these measures were meant to signal beyond diplomatic support for Georgia. While the presence of the destroyer may have implied a military threat, there were no public statements on the part of the United States to suggest that a U.S. intervention, no matter how limited, was under consideration. This does not mean that military options—including the bombing of the Roki Tunnel to prevent more Russian troops from pouring into South Ossetia—were not contemplated by some members of the administration—including, reportedly, Vice President Dick Cheney. However, while he listened to such options, President Bush appears to have dismissed them from the start out of concern that they could lead to a military confrontation with Moscow.[80]

For that same reason, the United States delegated most of the crisis resolution duties to the European Union. As Asmus notes, "Washington did not want to act unilaterally or allow this confrontation to become a U.S.-Russian fight that could spark a new cold war or even escalate into a military confrontation."[81] As a result, it was left to the European Union to fulfill U.S. and European objectives in Georgia, which broadly consisted of stopping the Russian offensive and preventing regime change.

The U.S. message also made it clear that Washington wanted to defuse the crisis, not escalate the conflict. Rice mentions an NSC meeting where the response to Russia was discussed:

> The session was a bit unruly, with a fair amount of chest beating about the Russians. At one point Steve Hadley intervened, something he rarely did. There was all kind of loose talk about what threats the United States might make. "I want to ask a question," he said in his low-key way. "Are we prepared to go to war with Russia over Georgia?" That quieted the room and we settled into a more productive conversation of what we could do.[82]

Rice also mentions the setup of a direct line between Chairman of the Joint Chiefs of Staff Mike Mullen and the Russian chief of staff. She describes the purpose of that line to be "largely to prevent any miscalculation between our forces."[83] Mullen,

---

[79] Labott, 2014.

[80] Asmus, 2010, pp. 186–187.

[81] Asmus, 2010, p. 177. At a Pentagon briefing, Defense Secretary Robert Gates stated, "The United States spent 45 years working very hard to avoid a military confrontation with Russia. I see no reason to change that approach today." Robert Gates, quoted in Steven Lee Myers and Thom Shanker, "Bush Aides Say Russia Actions in Georgia Jeopardize Ties," *New York Times*, August 14, 2008.

[82] Rice, 2011, p. 689.

[83] Rice, 2011, p. 689.

in particular, negotiated with Russia to bring back Georgian troops from Iraq into Georgia.[84]

Overall, U.S. threats to Russia following their invasion of Georgia were either vague—with Defense Secretary Robert Gates stating, for instance, "If Russia does not step back . . . the U.S.–Russian relationship could be adversely affected for years to come"[85]—or limited to moral condemnation and the threat of international isolation. Bush recalls telling Medvedev, "My strong advice is to start deescalating this thing now. . . . The disproportionality of your actions is going to turn the world against you. We're going to be with them."[86] He also told Putin that "he'd made a serious mistake and that Russia would isolate itself if it didn't get out of Georgia."[87] Overall, at no point did the United States threaten Moscow forcefully.

The Credibility of the Defender's Threats

The threat of international isolation was credible, as even Moscow's staunchest allies proved reluctant to support Russia's military intervention against Georgia, with the exception of Cuba and, after some hesitation (and some convincing by Moscow), Belarus.[88] While credible, how compelling was that threat for Russia? It is worth noting that Russia made clear efforts to justify its actions before the international community, arguing first that it acted in self-defense following the death of several of its peacekeepers in the Georgian attack; second, that its military reaction was little more than a reinforcement of its peacekeeping (turned "peace enforcement") presence; and third, that it intervened to stop the "genocide" of South Ossetians by Georgia, with Russian foreign minister Sergey Lavrov invoking the "responsibility to protect" as grounds for Russia's military intervention.[89] Yet Russia must have expected some degree of condemnation on the part of the international community once it crossed into Georgia's undisputed territories and accepted that cost—possibly

---

[84] Stent, 2014, p. 173, notes, "During the war, the Mullen-Makarov channel was the only working, high-level U.S.-Russian channel of communication, since Lavrov refused to talk to Rice for the first few days. They had seven conversations, and in one of them Mullen asked for Russian assistance in not hindering the return of Georgian troops from Iraq—which Makarov gave."

[85] Defense Secretary Robert Gates, quoted in Myers and Shanker, 2008.

[86] George W. Bush, *Decision Points*, New York: Crown Publishers, 2010, p. 434.

[87] Bush, 2010, p. 435.

[88] Oksana Antonenko, "A War with No Winner," *Survival*, Vol. 50, No. 5, October–November 2008, p. 27. Later in August, President Medvedev tried unsuccessfully to garner support from fellow members of the Shanghai Cooperation Organization. At their August 2008 meeting in Dushanbe, Tajikistan, they instead adopted a neutral position and reiterated their support to the principle of territorial integrity. See David L. Stern, "Security Group Refuses to Back Russia's Actions," *New York Times*, August 20, 2008.

[89] For more details on the Russian justifications for military intervention, and how they hold before international law, see Allison, 2009, pp. 1151–1157.

because, as Oksana Antonenko notes, Russia "does not see isolation as a problem, but rather as a tool of consolidation for its political elite."[90]

The Russian side appears to have missed the "signaling" mentioned by Hadley—which hinted at some possible military measures on the part of the United States—and the fact that the military option had been at some point on the table. An anecdote mentioned by Stent, based on an interview with Hadley, reveals, "In 2009, at a lunch with Russian ambassador Sergei Kislyak, the Russian asked, 'Did you really discuss sending U.S. troops to Georgia?' He was shocked when Hadley replied in the affirmative."[91] This anecdote suggests Russia (in so far as Kislyak's reaction was representative of the Russian leadership's beliefs) saw U.S. military involvement as unlikely, making it implausible that it would have influenced the Kremlin's decision to stop Russian troops. Yet more evidence from decisionmaking on the Russian side would be required to find out how U.S. messages impacted, if at all, decisions in Moscow.

Lessons

This second case presents an almost mirror image of the first one regarding the key variables outlined in Chapter Three (see Table E.2). While for U.S. efforts to deter Georgia, most variables were at a high level, in the case of U.S. efforts to deter Russia, they are all coded as low. The fact that Moscow initially wanted to get rid of Saakashvili but ended up not pushing its advantage when that objective was near (or at least closer than Moscow had ever been) suggests that Russia kept its objectives flexible and made an opportunistic decision not to take Tbilisi.

The United States chose to play a secondary role in the handling of the crisis in order to avoid a possible confrontation with Moscow, which explains why the U.S. message did not attempt to convey any clear or powerful threat. This points to the role that deterrence can play in crisis escalation—a role that was well understood by the Bush administration—as threats made, and particularly military threats, have either to be carried out with the potential of meeting a response in kind or risk being "empty threats" with broad international and domestic implications.[92]

---

[90] Antonenko, 2008, p. 28.

[91] Stent, 2014, p. 174.

[92] This argument is at the core of the audience cost theory first proposed by James Fearon in 1994. See James Fearon, "Domestic Political Audiences and the Escalation of International Disputes," *American Political Science Review*, Vol. 88, No. 3, September 1994, pp. 577–592. For a critique, see Jack Snyder and Erica D. Borghard, "The Cost of Empty Threats: A Penny, Not a Pound," *American Political Science Review*, Vol. 105, No. 3, August 2011, pp. 437–456.

Table E.2
Application of the Framework to the Case of the United States Deterring Russia

Category	Variable	Level in Present Case
How motivated was Russia?	General level of dissatisfaction with status quo and determination to create a new strategic situation.	Low. Russia was dissatisfied with Saakashvili's leadership, but its intervention had already changed the status quo and created a new strategic situation.
	Degree of fear that the strategic situation was about to turn against Russia in decisive ways.	Low. While the Russian offensive revealed major shortcomings of the armed forces, Russia was still dominating Georgian forces by far. The reaction of the international community was limited in scope and presented little risk for Russia.
	Level of national interest involved in specific territory of concern.	Low. While Tbilisi represents a strategic location and offered an opportunity for Russia to overthrow Saakashvili, Russia might have been more interested in the military bases that its forces found on their way to Tbilisi, which offered Russia an opportunity to damage Georgia's military capacity.
	Urgent sense of desperation or requirement to act; whether aggressor was locked into course of action.	Low. Russia had many options regarding where it could go, how fast it should get to Tbilisi, or whether it should go at all.
	Degree of aggressive, reckless, risk-accepting opportunism.	Low. There is no indication that Putin or Medvedev were ready to take overly risky actions.
	Level of motivated reasoning in play; degree of wishful thinking, misperception of basic strategic context.	Low. Nothing indicates that Russia did not have a good grasp of the strategic context and the opportunities and risks presented by various options.
Was the United States clear and explicit regarding what it sought to prevent and what actions it would take in response?	Precision in the type of aggression the United States sought to prevent.	High. The United States made clear it wanted the parties to cease fighting.
	Clarity in the actions that would be taken in the event of aggression.	Low. There were no real threats of actions the United States would take if Russia did not stop its forces.
	Forceful communication of these messages to outside audiences, especially potential aggressor(s).	Low. The United States mostly relied on the European Union to convey its message to Russia. Direct "threats" to Russia were limited and rather vague regarding what was being threatened.
	Timely response to warning with clarification of interests, threats.	Low. The United States position does not appear to have become more threatening as Russian forces were closing in on Tbilisi.
Did Russia view U.S. threats as credible and intimidating?	Actual and perceived strength of the local military capability to deny the presumed objectives of the aggression.	Low. U.S.-trained Georgian forces were quickly swept by the Russian offensive.
	Degree of automaticity of U.S. response, including escalation to larger conflict.	Low. Russia appears to have believed that the United States would not escalate militarily the conflict.

Table E.2—Continued

Category	Variable	Level in Present Case
	Degree of actual and perceived credibility of political commitment to fulfill deterrent threats.	Low. Deterrent threats were limited in the first place.
	Degree of national interests engaged in state to be protected.	Low. While Georgia was a U.S. ally, it was of limited strategic significance to the United States.
	Reputation for resolve with potential aggressor.	Low. While the United States and Russia had experienced severe tensions (e.g., U.S. support for color revolutions, U.S. missile defense project), they had never been in the situation of Russia invading a sovereign country.
	Degree of threat posed to aggressor's values and interests by the specific responses threatened by defender.	Low. U.S. responses were mostly confined to international isolation, which represents a limited threat to Russia's core values and interests.

Conclusion

The two cases of deterrence discussed in this appendix show a divergence between the quality of the deterrent message and the outcome of the deterrence effort. In the first case, the United States clearly expressed a credible threat message to Georgia, but Georgia went ahead with its attack against Tskhinvali anyway. In the second case, the U.S. deterrent message to Russia was weak by any measure, yet Russia did stop its advance toward Tbilisi, as the United States had hoped. In both cases, the factor that proved of critical importance to predict failure or success of deterrence was the degree of motivation of the aggressor. Georgia felt locked in a course of action, found itself with no good option, and went ahead with the attack regardless of what price it might have to pay for its decision. Moscow likely found that enough of its strategic and tactical objectives had been achieved by August 12, 2008, and that pushing into Tbilisi would be more trouble than it was worth.

The fact that the degree of aggressor's motivation was of paramount importance also means that in both cases, the outcome of U.S. deterrent efforts owed little to U.S. actions. Yet this does not mean that this outcome was entirely outside U.S. control. For instance, the United States could have taken some diplomatic steps earlier, as tensions escalated between Russia and Georgia in the months and years that preceded the August 2008 crisis. Extended deterrence against Russia at the time could have prevented Georgia from eventually finding itself in a situation where it saw a Russian military intervention as inevitable and military action as the only option. Extended deterrence would have come with its own costs, however, as it would have likely made U.S.-Russia discussions more difficult or conflictual on issues—such as missile defense—of greater strategic importance to the United States than the fate of Abkhazia and South Ossetia.

Unsurprisingly, it was easier for the United States to provide a clear and convincing deterrent message to its Georgian ally than to Russia. Saakashvili had been courting U.S. support, which he desperately needed to achieve his objectives of setting his country on a westward course and joining NATO. No matter what the United States did or did not do, it was unlikely to alienate Georgia—and even if it did, the strategic consequences would be minimal. Russia was a different story, and Washington was careful to steer away from anything that might look like, or trigger, a confrontation. While many of President Bush's decisions (particularly missile defense) antagonized Moscow, there were efforts, including as recently as April 2008 in Sochi, to reboot cooperation between the United States and Russia. As a result, Washington was exceedingly cautious in its messaging to Moscow. This prevented the United States from issuing a clear deterrent message and making threats that it did not want to deliver on.

Bibliography

"1967–1979: NATO's Readiness Increases," Supreme Headquarters Allied Powers Europe, n.d. As of May 22, 2018:
https://shape.nato.int/page14642550

Achen, Christopher H., and Duncan Snidal, "Rational Deterrence Theory and Comparative Case Studies," *World Politics*, Vol. 41, No. 2, January 1989, pp. 143–169.

Adamsky, Dmitry Dima, "The 1983 Nuclear Crisis–Lessons for Deterrence Theory and Practice," *Journal of Strategic Studies* Vol. 36, No. 1, 2013, pp. 4–41.

———, "Cross-Domain Coercion: The Current Russian Art of Strategy," Proliferation Papers 54, Paris: Institut français de relations internationales, November 2015.

Allison, Roy, "Russia Resurgent? Moscow's Campaign to 'Coerce Georgia to Peace,'" *International Affairs*, Vol. 84, No. 6, November 2009, pp. 1145–1171.

Antonenko, Oksana, "A War with No Winner," *Survival*, Vol. 50, No. 5, October–November 2008, pp. 23–36.

Arkin, William M., "Our Risky Naval Strategy Could Get Us All Killed," *Washington Post*, July 3, 1988. As of May 22, 2018:
https://www.washingtonpost.com/archive/opinions/1988/07/03/our-risky-naval-strategy-could-get-us-all-killed/717d1b1a-9679-4ddb-8673-1cd2690672f0/?utm_term=.8fc4a2775be6

Asmus, Ronald D., *Opening NATO's Door: How the Alliance Remade Itself for a New Era*, New York: Columbia University Press, 2002.

———, *A Little War That Shook the World: Georgia, Russia, and the Future of the West*, New York: St. Martin's Press, 2010.

Baldwin, David A., "The Power of Positive Sanctions," *World Politics*, Vol. 24, No. 1, October 1971, pp. 19–38. As of May 22, 1018:
https://pdfs.semanticscholar.org/21f1/275dfa6c659d02934435377c1f6cc59cb96e.pdf

Baker, James A., III, *The Politics of Diplomacy: Revolution, War and Peace, 1989–1992*, New York: G. P. Putnam's Sons, 1995.

Baker, Peter, "Trump's Previous View of NATO Is Now Obsolete," *New York Times*, April 13, 2017.

Bar, Shmuel, "Deterrence of Palestinian Terrorism: The Israeli Experience," in Andreas Wenger and Alex Wilner, eds., *Deterring Terrorism: Theory and Practice*, Stanford, Calif.: Stanford University Press, 2012.

Barlow, Jeffrey G., "NATO's Northern Flank: The Growing Soviet Threat," May 1, 1979, Heritage Foundation. As of May 22, 2018:
http://www.heritage.org/defense/report/natos-northern-flank-the-growing-soviet-threat

Barry, Ellen, "Russia President Dismisses Georgia's Leader as a 'Political Corpse,'" *New York Times*, September 2, 2008.

Batorshina, Irina, "Otnosheniie Pribaltiiskikh Respublik s Strategii Sderzhivaniia I Vovlecheniia Rossii (2014–2016)" [The attitude of the Baltic Republics to the strategy of containment and involvement of Russia (2014–2016)], *Problemy Natsional'noi Strategii*, Vol. 3, No. 42, 2017.

Beaufre, André, *Deterrence and Strategy*, New York: Praeger, 1965.

Beinart, Peter, "The U.S. Doesn't Need to Prove Itself in Ukraine," *Atlantic*, May 5, 2014.

Benson, Brett V., and Emerson M. Niou, "A Theory of Dual Deterrence: Credibility, Conditional Deterrence, and Strategic Ambiguity." Paper presented at the annual meeting of the Midwest Political Science Association, Chicago, April 12, 2007.

Berejikian, Jeffrey D., "A Cognitive Theory of Deterrence," *Journal of Peace Research*, Vol. 39, No. 2, March 2002, pp. 165–183. As of May 22, 2018:
http://www.jstor.org/stable/pdf/1555297.pdf

Berner, Örjan, *Soviet Policy Toward the Nordic Countries*, New York: University Press of America, 1986.

Betts, Richard K., *Nuclear Blackmail and Nuclear Balance*, Washington, D.C.: Brookings Institution Press, 2010.

Biden, Joseph, "Remarks by Vice President Joe Biden at the National Library of Latvia," August 24, 2016. As of May 24, 2018:
https://obamawhitehouse.archives.gov/the-press-office/2016/08/24/remarks-vice-president-joe-biden-national-library-latvia

———, "Remarks by the Vice President to Enhanced Forward Presence and Estonian Troops," July 31, 2017. As of May 24, 2018:
https://www.whitehouse.gov/briefings-statements/remarks-vice-president-enhanced-forward-presence-estonian-troops/

Biden, Joseph, Dahlia Grybauskaite, and Andris Berzins, "Remarks to the Press by Vice President Joe Biden, President Dahlia Grybauskaite of Lithuania, and President Andris Berzins of Latvia," March 19, 2014. As of May 24, 2018:
https://obamawhitehouse.archives.gov/the-press-office/2014/03/19/remarks-press-vice-president-joe-biden-president-dalia-grybauskaite-lith

Bitzinger, Richard A., *Assessing the Conventional Balance in Europe, 1945–1975*, Santa Monica, Calif.: RAND Corporation, N-2859-FF/RC, 1989. As of December 15, 2017:
https://www.rand.org/pubs/notes/N2859.html

Bjerga, Kjell Inge, *Politico-Military Assessments on the Northern Flank 1975–1990: Report from the IFS/PHP Bodø Conference of 20–21 August 2007*. As of May 22, 2018:
http://www.php.isn.ethz.ch/lory1.ethz.ch/documents/BodoeReport.pdf

Blackwill, Robert D., and Jeffrey W. Legro, "Constraining Ground Force Exercises of NATO and the Warsaw Pact," *International Security*, Vol. 14, No. 3, Winter 1989–1990, pp. 68–98.

Bohlen, Celestine, "Russia Cuts Gas Supply to Estonia in a Protest," *New York Times*, June 26, 1993. As of May 22, 2018:
http://www.nytimes.com/1993/06/26/world/russia-cuts-gas-supply-to-estonia-in-a-protest.html?mcubz=3

Børresen, Jacob, "Alliance Naval Strategies and Norway in the Final Years of the Cold War," *Naval War College Review*, Vol. 64, No. 2, Spring 2011.

Bound, Travis L., and Ryan C. Hendrickson, "Georgian Membership in NATO: Policy Implications of the Bucharest Summit," *Journal of Slavic Military Studies*, Vol. 22, No. 1, 2009, pp. 20–30.

Bowker, Mike, "The War in Georgia and the Western Response," *Central Asian Survey*, Vol. 30, No. 2, June 2011, pp. 197–211.

Brown, Michael E., *Deterrence Failures and Deterrence Strategies: Or, Did You Ever Have One of Those Days When No Deterrent Seemed Adequate?* Santa Monica, Calif.: RAND Corporation, P-5842, 1977. As of December 15, 2017:
http://www.rand.org/pubs/papers/P5842.html

Brundtland, Arne Olav, "The Nordic Balance: Past and Present," *Cooperation and Conflict*, Vol. 1, No. 4, 1966, pp. 30–63. As of May 22, 2018:
http://journals.sagepub.com/doi/pdf/10.1177/001083676600100403

Bueno de Mesquita, Bruce, *Principles of International Politics: People's Power, Preferences, and Perceptions*, Washington, D.C.: Congressional Quarterly Press, 2000.

Bueno de Mesquita, Bruce, James D. Morrow, and Ethan R. Zorick, "Capabilities, Perception, and Escalation," *American Political Science Review*, Vol. 91, No. 1, March 1997, pp. 15–27.

Bunn, M. Elaine, *Can Deterrence Be Tailored?* Strategic Forum No. 225, Washington, D.C.: National Defense University, Institute for National Strategic Studies, 2007.

Bureau of East Asian and Pacific Affairs, "U.S. Relations with New Zealand," U.S. Department of State, February 14, 2017. As of May 24, 2018:
https://www.state.gov/r/pa/ei/bgn/35852.htm

Burns, John F., "Andropov Offers Atom-Free Baltic," *New York Times*, June 7, 1983. As of May 24, 2018:
http://www.nytimes.com/1983/06/07/world/andropov-offers-atom-free-baltic.html

Bush, George W., *Decision Points*, New York: Crown Publishers, 2010.

Bush, Jason, and David Mardiste, "Russia and Estonia Swap Alleged Spies," Reuters, September 26, 2015. As of May 24, 2018:
http://www.reuters.com/article/us-russia-estonia-idUSKCN0RQ0GU20150926

Carlin, Wendy, "West German Growth and Institutions, 1945–1990," in Nicholas Crafts and Gianni Toniolo, eds., *Economic Growth in Europe Since 1945*, Cambridge: Cambridge University Press, 1996.

Carpenter, Ted Galen, "Are the Baltic States Next?" *National Interest*, March 24, 2014. As of May 24, 2018:
http://nationalinterest.org/commentary/are-the-baltic-states-next-10103

Carter, Donald A., *The U.S. Military Response to the 1960–1962 Berlin Crisis*, Washington, D.C.: U.S. Army Center of Military History, 2011. As of April 18, 2017:
https://www.archives.gov/files/research/foreign-policy/cold-war/1961-berlin-crisis/overview/us-military-response.pdf

Carter, Jimmy, "Proclamation 4648—30th Anniversary of NATO," March 22, 1979, American Presidency Project. As of May 22, 2018:
http://www.presidency.ucsb.edu/ws/index.php?pid=32079&st=norway&st1=

CDRSalamander, "Once More unto the Gap." As of May 22, 2018:
https://blog.usni.org/2016/04/20/once-more-unto-the-gap

Central Intelligence Agency, *Possibility of Direct Soviet Military Action During 1948*, Washington, D.C.: Central Intelligence Agency, March 30, 1948.

———, *Comprehensive Report of the Special Advisor to the DCI on Iraq's WMD*: Vol. 1, Washington, D.C.: Central Intelligence Agency, 2004.

Central Intelligence Agency Directorate of Intelligence, *Warsaw Pact: Planning for Operations Against Denmark*, April 1989. As of May 22, 2018:
https://www.cia.gov/library/readingroom/docs/1989-04-01.pdf

Chamberlain, Dianne Pfundstein, "NATO's Baltic Tripwire Forces Won't Stop Russia," *National Interest*, July 21, 2016. As of May 24, 2018:
http://nationalinterest.org/blog/the-skeptics/natos-baltic-tripwire-forces-wont-stop-russia-17074

Charap, Samuel, and Timothy J. Colton, *Everyone Loses: The Ukraine Crisis and the Ruinous Contest for Post-Soviet Eurasia*, Adelphi Series Vol. 56, London: International Institute for Strategic Studies, 2016.

Cheney, Dick, with Liz Cheney, *In My Time: A Personal and Political Memoir*, New York: Threshold Editions, 2011.

Chivers, C. J., "Georgia Offers Fresh Evidence on War's Start," *New York Times*, September 15, 2008.

Christiansson, Magnus, "Strategic Surprise in the Ukraine Crisis: Agendas, Expectations, and Organizational Dynamics in the EU Eastern Partnership Until the Annexation of Crimea 2014," master's thesis, Swedish National Defence College, August 2014.

Cigar, Norman, "Did Iraq Expect a Nuclear Desert Storm? Deterrence, Paradigms, and Operational Culture in a Weapons of Mass Destruction Environment," *War in History*, Vol. 21, No. 3, 2014, pp. 274–301.

Clemens, Walter C., and Franklyn Griffiths, "The Soviet Position on Arms Control and Disarmament—Negotiation and Propaganda, 1954–1964," Cambridge, Mass.: Center for International Studies, Massachusetts Institute of Technology, February 1, 1965, p. 52. As of May 22, 2018:
http://www.dtic.mil/dtic/tr/fulltext/u2/613518.pdf

Cockburn, Andrew, "The Bloom Comes Off the Georgian Rose," *Harper's*, October 31, 2013, As of May 24, 2018:
http://harpers.org/blog/2013/10/the-bloom-comes-off-the-georgian-rose/

———, "Game On: East vs. West, Again," *Harper's*, January 2015. As of May 24, 2018:
http://harpers.org/archive/2015/01/game-on/?single=1

Cohen, Ariel, and Robert F. Hamilton, *The Russian Military and the Georgian War: Lessons and Implications*, Carlisle, Penn.: Strategic Studies Institute, U.S. Army War College, June 2011.

Colby, Elbridge, "Russia's Evolving Nuclear Doctrine and Its Implications," Foundation for Strategic Research, January 2016. As of May 24, 2018:
https://www.frstrategie.org/publications/notes/russia-s-evolving-nuclear-doctrine-and-its-implications-2016-01

Collins, John M., "Principles of Deterrence," *Air University Review*, November–December 1979.

Congressional Budget Office, *The Marine Corps in the 1980s: Prestocking Proposals, the Rapid Deployment Force, and Other Issues*, Washington, D.C.: GPO, May 1980. As of May 22, 2018:
https://www.cbo.gov/sites/default/files/96th-congress-1979-1980/reports/80doc15.pdf

Conley, Heather, "The Baltic States in the World," Remarks to the Baltic American Freedom League, Los Angeles, April 24, 2004. As of May 24, 2018:
https://2001-2009.state.gov/p/eur/rls/rm/32363.htm

———, "Russia's Influence on Europe," in Craig Cohen and Josiane Gabel, eds., *2015 Global Forecast: Crisis and Opportunity*, Washington, D.C.: Center for Strategic and International Studies, 2014, pp. 28–32.

Conley, Heather A., Theodore P. Gerber, Lucy Moore, and Mihaela David, *Russian Soft Power in the 21st Century: An Examination of Russian Compatriot Policy in Estonia*, Washington D.C.: Center for Strategic and International Studies, August 2011.

Conley, Heather A., Kathleen H. Hicks, Lisa Sawyer Samp, Olga Oliker, John O'Grady, Jeffrey Rathke, Melissa Dalton, and Anthony Bell, *Evaluating Future U.S. Army Force Posture in Europe*, Washington, D.C.: Center for Strategic and International Studies, 2016.

Cooley, Alexander, "How the West Failed Georgia," *Current History*, October 2008, pp. 342–344.

Cooper, Helen, C. J. Chivers, and Clifford J. Levy, "U.S. Watched as a Squabble Turned into a Showdown," *New York Times*, August 17, 2008.

Cooper, Michael, and Elisabeth Bumiller, "War Puts Focus on McCain's Hard Line on Russia," *New York Times*, August 11, 2008.

Cornell, Svante E., "War in Georgia, Jitters All Around," *Current History*, October 2008, pp. 307–314.

Cornell, Svante E., and S. Frederick Starr, "Introduction," in Svante E. Cornell and S. Frederick Starr, eds., *The Guns of August 2008: Russia's War in Georgia*, Armonk, N.Y.: M.E. Sharpe 2009.

Council of the European Union, *Report of the Independent Fact-Finding Mission on the Conflict in Georgia*: Vol. I, Brussels: Council of the European Union, September 2009.

Crawford, Timothy W., "The Endurance of Extended Deterrence," in T. V. Paul, Patrick M. Morgan, and James J. Wirtz, eds., *Complex Deterrence: Strategy in the Global Age*, Chicago: University of Chicago Press, 2009.

Danilovic, Vesna, "Conceptual and Selection Bias Issues in Deterrence," *Journal of Conflict Resolution*, Vol. 45, No. 1, February 2001a, pp. 97–125.

———, "The Sources of Threat Credibility in Extended Deterrence," *Journal of Conflict Resolution*, Vol. 45, No. 3, June 2001b, pp. 341–369.

———, "Deterrence and Conflict," in *When the Stakes Are High: Deterrence and Conflict Among Major Powers*, Ann Arbor: University of Michigan Press, 2002, pp. 47–68. As of May 22, 2018:
https://www.press.umich.edu/pdf/0472112872-ch3.pdf

Defense Manpower Data Center, "Worldwide Manpower Distribution by Geographical Area (M05): Historical Reports—Military Only, 1950, 1953–1999," n.d. As of June 8, 2018:
http://digital-commons.usnwc.edu/nwc-review/vol64/iss2/7/

———, *Historical Report—Military Only* (aggregated data 1950–current), Alexandria, Va.: Department of Defense, 2017. As of April 2017:
https://www.dmdc.osd.mil/appj/dwp/dwp_reports.jsp

"Defense of Greenland: Agreement Between the United States and the Kingdom of Denmark, April 27, 1951," Avalon Project, Lillian Goldman Law Library, Yale Law School, April 27, 1951. As of May 22, 2018:
http://avalon.law.yale.edu/20th_century/den001.asp

Delpech, Thérèse, *Nuclear Deterrence in the 21st Century: Lessons from the Cold War for a New Era of Strategic Piracy*, Santa Monica, Calif.: RAND Corporation, MG-1103-RC, 2012. As of December 15, 2017:
https://www.rand.org/pubs/monographs/MG1103.html

Doeker, Gunther, Klaus Melsheimer, and Dieter Schroder, "Berlin and the Quadripartite Agreement of 1971," *American Journal of International Law*, Vol. 67, No. 1, January 1973, pp. 44–62.

Downie, Leonard, Jr., "Scandinavia, Alarmed by Afghanistan, Reviews Its Defenses," *Washington Post*, February 5, 1980. As of May 22, 2018:
https://www.washingtonpost.com/archive/politics/1980/02/05/scandinavia-alarmed-by-afghanistan-reviews-its-defenses/d998dfca-6cd8-44cb-a4c8-20aa619ac5f1/?utm_term=.eebf93044840

Duelfer, Charles A., and Stephen Benedict Dyson, "Chronic Misperception and International Conflict: The U.S.-Iraq Experience," *International Security*, Vol. 36, No. 1, Summer 2011, pp. 73–100.

Dyndal, Gjert Lage, "How the High North Became Central in NATO Strategy: Revelations from the NATO Archives," *Journal of Strategic Studies*, Vol. 34, No. 4, August 2011, pp. 557–585. As of May 22, 2018:
http://www.tandfonline.com/doi/pdf/10.1080/01402390.2011.561094?needAccess=true

———, "50 Years Ago: The Origins of NATO Concerns About the Threat of Russian Strategic Nuclear Submarines," *NATO Review*, March 24, 2017. As of May 22, 2018:
http://www.nato.int/docu/review/2017/Also-in-2017/50-years-ago-nato-concerns-threat-russian-strategic-nuclear-submarines-soviet-bastion-high-north/EN/index.htm

Dzhindzhikhashvili, Misha, "Georgia: Russian Jet Fired Missile," *Washington Post*, August 8, 2007. As of May 25, 2018:
http://www.washingtonpost.com/wp-dyn/content/article/2007/08/08/AR2007080800336_pf.html

Ehala, Martin, "The Bronze Soldier: Identity Threat and Maintenance in Estonia," *Journal of Baltic Studies*, Vol. 40, No. 1, March 2009, pp. 139–158.

Emmott, Robin, and Sabine Siebold, "NATO Agrees to Reinforce Eastern Poland, Baltic States Against Russia," Reuters, July 7, 2016. As of May 24, 2018:
https://www.reuters.com/article/us-nato-summit/nato-agrees-to-reinforce-eastern-poland-baltic-states-against-russia-idUSKCN0ZN2NL

Erickson, John, "The Northern Theater: Soviet Capabilities and Concepts," *Strategic Review*, Vol. 4, No. 3, Summer 1976.

Fearon, James, "Domestic Political Audiences and the Escalation of International Disputes," *American Political Science Review*, Vol. 88, No. 3, September 1994, pp. 577–592.

Felgenhauer, Pavel, "Moscow Ready for Major Confrontations with Pro-Western Georgia and Ukraine," Jamestown Foundation, *Eurasia Daily Monitor*, Vol. 5, No. 117, June 19, 2008. As of May 25, 2018:
https://jamestown.org/program/moscow-ready-for-major-confrontations-with-pro-western-georgia-and-ukraine/

Ferdinando, Lisa, "Russian Airspace Violations in Nordic-Baltic Regions Dangerous, Work Says," U.S. Department of Defense, October 6, 2016. As of May 24, 2018:
https://www.defense.gov/News/Article/Article/968780/russian-airspace-violations-in-nordic-baltic-region-dangerous-work-says/

Fisher, Max, "Donald Trump's Ambivalence on the Baltics Is More Important Than It Seems," *New York Times*, July 21, 2016.

Fletcher, Brooks, "Saber Strike 2013 a Demonstration of Multinational Partnership in the Baltics," U.S. Army Europe Public Affairs, June 13, 2013. As of May 24, 2018:
https://media.defense.gov/2018/May/02/2001911402/-1/-1/0/06132013%20SABER%20STRIKE%202013%20A%20DEMONSTRATION%20OF%20MULTINATIONAL%20PARTNERSHIP%20IN%20THE%20BALTICS.PDF

Ford, Gerald, "Text of an Address Before the Council of the North Atlantic Treaty Organization in Brussels," May 29, 1975, American Presidency Project. As of May 22, 2018:
http://www.presidency.ucsb.edu/ws/index.php?pid=4948&st=article+5&st1=

Frederick, Bryan, Matthew Povlock, Stephen Watts, Miranda Priebe, and Edward Geist, *Assessing Russian Reactions to U.S. and NATO Posture Enhancements*, Santa Monica, Calif.: RAND Corporation, RR-1879-AF, 2017. As of May 24, 2018:
https://www.rand.org/pubs/research_reports/RR1879.html

Frederick, Bryan A., Paul R. Hensel, and Christopher Macaulay, "The Issue Correlates of War Territorial Claims Data, 1816–2011," *Journal of Peace Research*, Vol. 54, No. 1, 2017, pp. 99–108.

Freedman, Lawrence, *Strategic Coercion: Concepts and Cases*, Oxford: Oxford University Press, 1998.

———, *Deterrence*, Cambridge: Polity Press, 2004.

Fuhrmann, Matthew, and Sarah E. Kreps, "Targeting Nuclear Programs in War and Peace: A Quantitative Empirical Analysis, 1941–2000," *Journal of Conflict Resolution*, Vol. 54, No. 6, December 2010, pp. 831–859.

Fuhrmann, Matthew, Matthew Kroenig, and Todd S. Sechser, "Response: The Case for Using Statistics to Study Nuclear Security," *H-Diplo/ISSF Forum*, No. 2, 2014, pp. 37–54.

Fulcher, Kara Stibora, "A Sustainable Position? The United States, the Federal Republic, and the Ossification of Allied Policy on Germany, 1958–1962," *Diplomatic History*, Vol 26, No. 2, 2002, pp. 283–307.

Gaddis, John Lewis, *The Cold War: A New History*, New York: Penguin Books, 2006.

Gallis, Paul, *The NATO Summit at Bucharest, 2008*, Washington, D.C.: Congressional Research Service, May 5, 2008.

"The Gates Hearings; Early Indicators of Kuwait Invasion." *New York Times*, September 25, 1991. As of May 24, 2018:
http://www.nytimes.com/1991/09/25/us/the-gates-hearings-early-indicators-of-kuwait-invasion.html

Gause, F. Gregory, III, "Iraq and the Gulf War: Decision-Making in Baghdad." Unpublished manuscript, n.d. As of June 6, 2018:
https://www.files.ethz.ch/isn/6844/doc_6846_290_en.pdf

Gavin, Francis J., "What We Talk About When We Talk About Nuclear Weapons: A Review Essay," *H-Diplo/ISSF Forum*, No. 2, 2014, pp. 11–36.

Gelb, Leslie H., "Mr. Bush's Fateful Blunder." *New York Times*, July 17 1991. As of May 24, 2018:
http://www.nytimes.com/1991/07/17/opinion/foreign-affairs-mr-bush-s-fateful-blunder.html

Gelb, Norman, *The Berlin Wall: Kennedy, Khrushchev, and a Showdown in the Heart of Europe*, New York: Times Books, 1986.

George, Alexander L., "The General Theory and Logic of Coercive Diplomacy," in Security, Strategy, and Forces Faculty, eds., *Strategy and Force Planning*, 2nd ed., Newport, R.I.: Naval War College Press, 1991.

George, Alexander L., and William E. Simons, *The Limits of Coercive Diplomacy*, 2nd ed., Boulder, Colo.: Westview Press, 1994.

George, Alexander L., and Richard Smoke, *Deterrence in American Foreign Policy: Theory and Practice*, New York: Columbia University Press, 1974.

———, "Deterrence and Foreign Policy," *World Politics*, Vol. 41, No. 2, January 1989, pp. 170–182.

German, Robert K., "Norway and the Bear: Soviet Coercive Diplomacy and Norwegian Security Policy," *International Security*, Vol. 7, No. 2, Fall 1982. As of May 22, 2018:
http://www.jstor.org/stable/pdf/2538433.pdf?refreqid=excelsior%3A63d2343eb7c64adbd59c21eb3e5934be

Ghosn, Faten, and Scott Bennett, *Codebook for the Dyadic Militarized Interstate Incident Data*, Version 3.10, 2003. As of May 24, 2018:
http://cow.la.psu.edu/COW2%20Data/MIDs/Codebook%20for%20Dyadic%20MID%20Data%20v3.10.pdf

Golubev, Alexey, "*Kholodnaya voina v Arktike* (Review)," *Journal of Cold War Studies*, Vol. 14, No. 1, Winter 2012, pp. 143–147. As of May 22, 2018:
https://muse.jhu.edu/article/467700/pdf

Gordon, Michael R. "Denmark Agrees to Nuclear Policy," *New York Times*, June 8, 1988, p. A14.

———, "U.S. War Game in West Germany to Be Cut Back," *New York Times*, December 14, 1989. As of May 24, 2018:
http://www.nytimes.com/1989/12/14/world/us-war-game-in-west-germany-to-be-cut-back.html

———, "U.S. Deploys Air and Sea Forces After Iraq Threatens 2 Neighbors," *New York Times*, July 25, 1990a. As of June 8, 2018:
http://www.nytimes.com/1990/07/25/world/us-deploys-air-and-sea-forces-after-iraq-threatens-2-neighbors.html

———, "Iraq Army Invades Capital of Kuwait in Fierce Fighting," *New York Times*, August 2, 1990b. As of June 8, 2018:
http://www.nytimes.com/1990/08/02/world/iraq-army-invades-capital-of-kuwait-in-fierce-fighting.html

———, "Pentagon Objected to Bush's Message to Iraq," *New York Times*, October 25, 1992. As of May 24, 2018:
http://www.nytimes.com/1992/10/25/world/pentagon-objected-to-bush-s-message-to-iraq.html?pagewanted=all

Gordon, Michael R., and Bernard E. Trainor, *The Generals' War: The Inside Story of the Conflict in the Gulf*, New York: Little, Brown and Company, 1995.

Gorenburg, Dmitry, "Countering Color Revolutions: Russia's New Security Strategy and Its Implications for U.S. Policy," *Russian Military Reform*, September 15, 2014. As of May 24, 2018:
https://russiamil.wordpress.com/2014/09/15/countering-color-revolutions-russias-new-security-strategy-and-its-implications-for-u-s-policy/

Grathwol, Robert P., and Donita M. Moorhus, *American Forces in Berlin: Cold War Outpost, 1945–1994*, Washington, D.C.: U.S. Department of Defense, Legacy Resource Management Program, 1995.

Grigas, Agnia, "Energy Policy: The Achilles Heel of the Baltic States," in Agnia Grigas, Andres Kasekamp, Kristina Maslauskaite, and Liva Zorgenfreija, *The Baltic States in the EU: Yesterday, Today and Tomorrow*, Studies and Reports No. 98, Paris: Notre Europe/Jacques Delors Institute, July 2013.

Grushko, Alexander, "Speech by Russia's Permanent Representative to NATO Alexander Grushko at the opening of the OSCE Annual Security Review Conference (ASRC) in Vienna," June 27, 2017. As of May 24, 2018:
http://www.mid.ru/en/foreign_policy/rso/nato/-/asset_publisher/ObVB8wSP5tE2/content/id/2799459

Haass, Richard N., *War of Necessity, War of Choice: A Memoir of Two Iraq Wars*, New York: Simon and Schuster, 2009

Hamm, Manfred R., "Ten Steps to Counter Moscow's Threat to Northern Europe," Backgrounder No. 356, Washington, D.C.: Heritage Foundation, May 30, 1984. As of May 22, 2018:
https://www.heritage.org/europe/report/ten-steps-counter-moscows-threat-northern-europe

Hanna, Andrew, "How a Tiny Baltic Nation Became a Top Destination for U.S. Officials," *Politico*, July 29, 2017. As of May 24, 2018:
http://www.politico.com/story/2017/07/29/estonia-russia-us-visits-pence-241098

Harrington, Daniel F., *Berlin on the Brink: The Blockade, the Airlift, and the Early Cold War*, Lexington: University Press of Kentucky, 2012.

Harvey, Frank P., "Rational Deterrence Theory Revisited: A Progress Report," *Canadian Journal of Political Science*, Vol. 28, No. 3, 1995, pp. 403–436.

Hattendorf, John B., and Peter M. Swartz, eds., *U.S. Naval Strategy in the 1980s: Selected Documents*, Newport, R.I.: Naval War College Press, 2008. As of May 22, 2018:
https://fas.org/irp/doddir/navy/strategy1980s.pdf

Helmer-Hirschberg, Olaf, *Deterrence*, Santa Monica, Calif.: RAND Corporation, RM-1882, 1957. As of December 15, 2017:
http://www.rand.org/pubs/research_memoranda/RM1882.html

Herb, Jeremy, "Trump Commits to NATO's Article 5," *CNN*, June 9, 2017. As of May 24, 2018:
https://www.cnn.com/2017/06/09/politics/trump-commits-to-natos-article-5/index.html

Hermes, Walter G., "Global Pressures and the Flexible Response," in William A. Stofft, ed., *American Military History*, Washington D.C.: U.S. Army Center of Military History, 1989, pp. 591–619. As of May 22, 2018:
http://www.history.army.mil/books/AMH/AMH-27.htm

Herszenhorn, David M., "Russia and Estonia Differ over Detention," *New York Times*, September 5, 2014. As of May 24, 2018:
https://www.nytimes.com/2014/09/06/world/europe/russia-detains-estonian-officer-raising-tensions.html

Hill, Fiona, and Clifford G. Gaddy, *Mr. Putin: Operative in the Kremlin*, Washington, D.C.: Brookings Institution Press, 2013.

Holbraad, Carsten, "Denmark: Half-Hearted Partner," in Nils Ørvik, ed., *Semialignment and Western Security*, New York: St. Martin's Press, 1986.

Holmes, Steven A., "Congress Backs Curbs Against Iraq," *New York Times*, July 28, 1990. As of May 24, 2018:
http://www.nytimes.com/1990/07/28/world/congress-backs-curbs-against-iraq.html

Holst, Johan J., "Norway's Search for a Nordpolitik," *Foreign Affairs*, Fall 1981. As of May 22, 2018:
https://www.foreignaffairs.com/articles/norway/1981-09-01/norways-search-nordpolitik

Hooker, Richard D., Jr., "NATO's Northern Flank: A Critique of the Maritime Strategy," *Parameters*, June 1989. As of May 22, 2018:
http://ssi.armywarcollege.edu/pubs/parameters/Articles/1989/1989%20hooker%20nato.pdf

Hopf, Theodore G., *Peripheral Visions: Deterrence Theory and American Foreign Policy in the Third World, 1965–1990*, Ann Arbor: University of Michigan Press, 1994.

Hotta, Eri, *Japan 1941: Countdown to Infamy*, New York: Alfred A. Knopf, 2013.

Human Rights Watch, "Georgian Villages in South Ossetia Burnt, Looted," August 12, 2008. As of June 8, 2018:
https://www.hrw.org/news/2008/08/12/georgian-villages-south-ossetia-burnt-looted

Huth, Paul, and Bruce Russett, "What Makes Deterrence Work? Cases from 1900 to 1980," *World Politics*, Vol. 36, No. 4, July 1984, pp. 496–526.

———, "Deterrence Failure and Crisis Escalation," *International Studies Quarterly*, Vol. 32, No. 1, March 1988, pp. 29–45.

———, "Testing Deterrence Theory: Rigor Makes a Difference," *World Politics*, Vol. 42, No. 4, July 1990, pp. 466–501.

———, "General Deterrence Between Enduring Rivals: Testing Three Competing Models," *American Political Science Review*, Vol. 87, No. 1, March 1993, pp. 61–73.

Huth, Paul K., "Extended Deterrence and the Outbreak of War," *American Political Science Review*, Vol. 82, No. 2, June 1988, pp. 423–443.

———, *Extended Deterrence and the Prevention of War*, New Haven, Conn.: Yale University Press, 1988.

———, "Deterrence and International Conflict: Empirical Findings and Theoretical Debates," *Annual Review of Political Science*, Vol. 2, 1999, pp. 25–48.

Huth, Paul K., Christopher Gelpi, and D. Scott Bennett, "The Escalation of Great Power Militarized Disputes: Testing Rational Deterrence Theory and Structural Realism," *American Political Science Review*, Vol. 87, No. 3, September 1993, pp. 609–623.

Ibrahim, Youssef, "Iraq Said to Prevail in Oil Dispute with Kuwait and Arab Emirates," *New York Times*, July 26, 1990. As of May 24, 2018:
http://www.nytimes.com/1990/07/26/world/iraq-said-to-prevail-in-oil-dispute-with-kuwait-and-arab-emirates.html

Illarionov, Andrei, "The Russian Leadership Preparation for War," in Svante E. Cornell and S. Frederick Starr, eds., *The Guns of August 2008: Russia's War in Georgia*, Armonk, N.Y.: M.E. Sharpe, 2009.

"Information Memorandum from the Assistant Secretary of State for European Affairs (Hillenbrand) to Secretary of State Rogers," in James E. Miller and Laurie Van Hook, eds., *Foreign Relations of the United States, 1969–1976*: Vol. XLI, *Western Europe; NATO, 1969–1972*, Washington, D.C.: GPO, 2012. As of May 22, 2018:
https://s3.amazonaws.com/static.history.state.gov/frus/frus1969-76v41/pdf/frus1969-76v41.pdf

International Crisis Group, *Saakashvili's Ajara Success: Repeatable Elsewhere in Georgia?* Brussels: International Crisis Group, August 18, 2004.

———, *Georgia: Sliding Towards Authoritarianism?* Europe Report No. 189, Brussels: International Crisis Group, December 19, 2007.

"Inverviu Vladimira Putina frantsuzkoi gazete Le Figaro" [Vladimir Putin's interview with the French newspaper Le Figaro], May 31, 2017. As of May 24, 2018:
http://kremlin.ru/events/president/news/54638

Ioffe, Julia, "How Russia Saw the 'Red Line' Crisis," *Atlantic*, March 11, 2016.

Janes, Robert W., "The Soviet Union and Northern Europe: New Thinking and Old Security Constraints," *Annals of the American Academy of Political and Social Science*, Vol. 512, No. 1, November 1990, pp. 163–172. As of May 22, 2018:
http://www.jstor.org/stable/pdf/1046895.pdf?refreqid=excelsior:6398adfc04619123138b48ae27858ef4

Jervis, Robert, "Deterrence Theory Revisited: Review Article," *World Politics*, Vol. 31, No. 2, January 1979, pp. 289–324. As of May 22, 2018:
https://www.jstor.org/stable/pdf/2009945.pdf

———, "Deterrence and Perception," *International Security*, Vol. 7, No. 3, Winter 1982–1983, pp. 3–30.

———, "Rational Deterrence: Theory and Evidence," *World Politics*, Vol. 41, No. 2. January 1989, pp. 183–207.

———, "The Confrontation Between Iraq and the US: Implications for the Theory and Practice of Deterrence," *European Journal of International Relations*, Vol. 9, No. 2, September 2003, pp. 315–337.

Jervis, Robert, Richard Ned Lebow, and Janice Gross Stein, *Psychology and Deterrence*, Baltimore: Johns Hopkins University Press, 1985.

Johnson, David E., Karl P. Mueller, and William H. Taft, *Conventional Coercion Across the Spectrum of Operations: The Utility of U.S. Military Forces in the Emerging Security Environment*, Santa Monica, Calif.: RAND Corporation, MR-1494-A, 2002. As of December 15, 2017:
https://www.rand.org/pubs/monograph_reports/MR1494.html

Johnson, Jesse C., Brett Ashley Leeds, and Ahra Wu, "Capability, Credibility, and Extended General Deterrence," *International Interactions*, Vol. 41, No. 2, 2015, pp. 309–336.

Johnson, Lyndon B., "Remarks to Members of the NATO Parliamentarians Conference," September 18, 1964, American Presidency Project. As of May 22, 2018:
http://www.presidency.ucsb.edu/ws/index.php?pid=26512&st=nato&st1=

"Joint Military Exercises Begin This Week," UPI, August 29, 1982. As of May 22, 2018:
https://www.upi.com/Archives/1982/08/29/Joint-military-exercises-begin-this-week/5082399441600/

Jones, John, "A Naval Force of 35,000 Personnel and 150 Ships . . ." UPI, August 28, 1986. As of May 22, 2018:
http://www.upi.com/Archives/1986/08/28/A-naval-force-of-35000-personnel-and-150-ships/5006525585600/

Judt, Tony, *Postwar: A History of Europe Since 1945*. New York: Penguin Press, 2005.

Kagan, Robert, "Putin Makes His Move," *Washington Post*, August 11, 2008.

Kang, Kyungkook, and Jacek Kugler, "Conditional Deterrence: Parity, Terrorism and Tenuous Deterrence," *Terrorism and Tenuous Deterrence*, January 7, 2015.

Karabell, Zachary, "Backfire: US Policy Toward Iraq, 1988–2 August 1990," *Middle East Journal*, Vol. 49, No. 1, 1995, pp. 28–47.

Karsh, Efraim, and Inari Rautsi, *Saddam Hussein: A Political Biography*, New York: The Free Press, 1991.

Kasekamp, Andres, *A History of the Baltic States*, Houndsmills, England: Palgrave Macmillan, 2010.

Kempe, Frederick. *Berlin 1961: Kennedy, Khrushchev, and the Most Dangerous Place on Earth*, New York: G. P. Putnam's Sons, 2011.

Kennedy, John F., "Joint Statement Following Discussions with Prime Minister Gerhardsen of Norway," May 11, 1962, American Presidency Project. As of May 22, 2018:
http://www.presidency.ucsb.edu/ws/index.php?pid=8647&st=norway&st1=

———, "Remarks in Naples at NATO Headquarters," July 2, 1963, American Presidency Project. As of May 22, 2018:
http://www.presidency.ucsb.edu/ws/index.php?pid=9332&st=nato&st1=

Kernan, Robert F., *Norway and the Northern Front: Wartime Prospects*, Maxwell Air Force Base, Ala.: Air War College, May 1989. As of May 22, 2018:
http://www.dtic.mil/dtic/tr/fulltext/u2/a217547.pdf

Khrulev, Lieutenant General Anatoly, interview, April 8, 2014, translated from Russian by Samuel Charap. As of May 25, 2018:
http://www.kp.ru/daily/26266/3144362/

Kimsey, Doug, "Training for Iraq Boosts Security in the Caucasus," June 28, 2005, U.S. Department of Defense. As of May 24, 2018:
http://archive.defense.gov/news/newsarticle.aspx?id=16284

Knopf, Jeffrey W., "Three Items in One: Deterrence as Concept, Research Program, and Political Issue," in T. V. Paul, Patrick M. Morgan, and James J. Wirtz, eds., *Complex Deterrence: Strategy in the Global Age*, Chicago: University of Chicago Press, 2009.

Kofman, Michael, "Russian Military Buildup in the West: Fact Versus Fiction," *Russia Matters*, September 7, 2017. As of May 24, 2018:
https://www.russiamatters.org/analysis/russian-military-buildup-west-fact-versus-fiction

Kofman, Michael, Katya Migacheva, Brian Nichiporuk, Andrew Radin, Olesya Tkacheva, and Jenny Oberholtzer, *Lessons from Russia's Operations in Crimea and Eastern Ukraine*, Santa Monica, Calif.: RAND Corporation, RR-1498-A, 2017. As of May 24, 2018:
https://www.rand.org/pubs/research_reports/RR1498.html

Kramer, Franklin D., and Bantz J. Craddock, *Effective Defense of the Baltics*, Washington D.C.: Atlantic Council, 2016.

Kroenig, Matthew, "Nuclear Superiority and the Balance of Resolve: Explaining Nuclear Crisis Outcomes," *International Organization*, Vol. 67, No. 1, January 2013, pp. 141–171.

Kugler, Richard L., *The Great Strategy Debate: NATO's Evolution in the 1960s*, Santa Monica, Calif.: RAND Corporation, N-3252-FF/RC, 1991. As of December 15, 2017:
https://www.rand.org/pubs/notes/N3252.html

Labott, Elise, "Stephen J. Hadley Looks Back on 9/11, Iraq, and Afghanistan," October 22, 2014, Council on Foreign Relations. As of May 26, 2018:
http://www.cfr.org/politics-and-strategy/stephen-j-hadley-looks-back-911-iraq-afghanistan/p35753

Larrabee, F. Stephen, *The Baltic State and NATO Membership*, Santa Monica, Calif.: RAND Corporation, CT-204, 2003. As of May 24, 2018:
https://www.rand.org/pubs/testimonies/CT204.html

Lebow, Richard Ned, *Between Peace and War: The Nature of International Crisis*, Baltimore: John Hopkins University Press, 1981.

———, "Misconceptions in American Strategic Assessment," *Political Science Quarterly*, Vol. 97, No. 2, Summer 1982.

———, "The Deterrence Deadlock: Is There a Way Out?" *Political Psychology*, Vol. 4, No. 2, June 1983, pp. 333–354.

———, "Windows of Opportunity: Do States Jump Through Them?" *International Security*, Vol. 9, No. 1, Summer 1984, pp. 147–186.

———, "Deterrence Failure Revisited," *International Security*, Vol. 12, No. 1, Summer 1987, pp. 197–213.

———, "Deterrence and Reassurance: Lessons from the Cold War," *Global Dialogue*, Autumn 2001, pp. 119–132.

———, "Thucydides and Deterrence," *Security Studies* Vol. 16, No. 2, April–June 2007, pp. 163–188. As of May 24, 2018:
http://www.dartmouth.edu/~nedlebow/thuc_deterr.pdf

Lebow, Richard Ned, and Janice Gross Stein, "Rational Deterrence Theory: I Think, Therefore I Deter," *World Politics*, Vol. 41, No. 2, January 1989, pp. 208–224. As of May 24, 2018:
https://www.jstor.org/stable/pdf/2010408.pdf

———, "Deterrence: The Elusive Dependent Variable," *World Politics*, Vol. 42, No. 3, April 1990, pp. 336–369. As of May 22, 2018:
https://www.jstor.org/stable/pdf/2010415.pdf

———, "Deterrence and the Cold War," *Political Science Quarterly*, Vol. 110, No. 2, Summer 1995, pp. 157–181.

Levada-Center, "Presidential Election," press release, May 29, 2017. As of May 24, 2018:
https://www.levada.ru/en/2017/05/29/presidential-election/

Levy, Jack S., "When Do Deterrent Threats Work?" *British Journal of Political Science*, Vol. 18, No. 4, October 1988, pp. 485–512.

———, "Deterrence and Coercive Diplomacy: The Contributions of Alexander George," *Political Psychology*, Vol. 29, No. 4, August 2008, pp. 537–552.

Long, Austin, *Deterrence, from Cold War to Long War: Lessons from Six Decades of RAND Research*, Santa Monica, Calif.: RAND Corporation, MG-636-OSD/AF, 2008. As of July 19, 2018:
https://www.rand.org/pubs/monographs/MG636.html

Long, Duncan, Terrence K. Kelly, and David C. Gompert, eds., *Smarter Power, Stronger Partners*: Vol. II, *Trends in Force Projection Against Potential Adversaries*, Santa Monica, Calif.: RAND Corporation, RR-1359/1-A, 2017. As of May 22, 2018:
https://www.rand.org/pubs/research_reports/RR1359z1.html

Lukacs, John, "Finland Vindicated," *Foreign Affairs*, Fall 1992. As of May 22, 2018:
https://www.foreignaffairs.com/articles/russia-fsu/1992-09-01/finland-vindicated

Lukyanov, Fyodor, "Stanet li Polsha modelyu dlya Baltii?" [Will Poland become the model for the Baltics?], *Russia in Global Affairs*, April 18, 2012.

Lunák, Petr, "Khrushchev and the Berlin Crisis: Soviet Brinkmanship Seen from Inside," *Cold War History*, Vol. 3, No. 2, January 2003, pp. 53–82.

Lupovici, Amir, "The Emerging Fourth Wave of Deterrence Theory—Toward a New Research Agenda," *International Studies Quarterly*, Vol. 54, No. 3, September 2010, pp. 705–732.

Mayhugh, Tryphena, "U.S., NATO Conclude Saber Strike 17 Exercise," U.S. Department of Defense, June 26, 2017. As of May 24, 2018:
https://www.defense.gov/News/Article/Article/1229124/

McCormick, Gordon H., *Stranger Than Fiction: Soviet Submarine Operations in Swedish Waters*, Santa Monica, Calif.: RAND Corporation, R-3776-AF, January 1990. As of May 25, 2018:
https://www.rand.org/pubs/reports/R3776.html

McCurry, Justin, "Trump Says U.S. May Not Automatically Defend NATO Allies Under Attack," *Guardian*, July 21, 2016.

McDermott, Roger N., "Russia's Conventional Armed Forces and the Georgian War," *Parameters*, Spring 2009, pp. 67–73.

McNamara, Eoin Micheál, "Securing the Nordic-Baltic Region," *NATO Review Magazine*, n.d. As of May 24, 2018:
http://www.nato.int/docu/review/2016/Also-in-2016/security-baltic-defense-nato/EN/index.htm

Mearsheimer, John J., *Conventional Deterrence*, Ithaca, N.Y.: Cornell University Press, 1983.

Medvedev, Dmitry, interview, August 5, 2011, translated from Russian by Samuel Charap. As of May 25, 2018:
http://echo.msk.ru/programs/beseda/799478-echo/

"Memorandum of Conversation," January 29, 1979, in Adam M. Howard, ed., *Foreign Relations of the United States, 1977–1980*: Vol. XIII, *China*, Washington, D.C.: GPO, 2013, p. 747. As of May 22, 2018:
https://history.state.gov/historicaldocuments/frus1977-80v13/d203

Menkiszak, Marek, *Russia's Best Enemy: Russian Policy Towards the United States in Putin's Era*, Point of View No. 62, Warsaw: Center for Eastern Studies, February 2017.

Merabishvili, Vano, interview, October 29, 2010. As of May 24, 2018:
https://www.kommersant.ru/doc/1048935

Mercer, Jonathan, *Reputation and International Politics*, Ithaca, N.Y.: Cornell University Press, 1996.

Mihkelson, Marko, "Russia's Policy Toward Ukraine, Belarus, Moldova, and the Baltic States," in Janusz Bugajski, ed., *Toward an Understanding of Russia: New European Perspectives*, New York: Council on Foreign Relations, 2002.

Milburn, Thomas W., "What Constitutes Effective Deterrence?" *Conflict Resolution*, Vol. 3, No. 2, June 1959, pp. 138–145. As of May 22, 2018:
https://www.jstor.org/stable/pdf/173109.pdf

Miller, Roger G., *To Save a City: The Berlin Airlift, 1948–1949*, Washington, D.C.: Air Force History and Museums Program, 1998.

"Mir bez Illyuziy i Mifov" [Peace without illusions or myths], *Rossiiskaya Gazeta*, January 15, 2017.

Morgan, Clifton, Navin Bapat, and Yoshi Kobayashi, "The Threat and Imposition of Sanctions: Updating the TIES Dataset." *Conflict Management and Peace Science*, Vol. 31, No. 5, 2014, pp. 541–558.

Morgan, Forrest E., *Compellence and the Strategic Culture of Imperial Japan: Implications for Coercive Diplomacy in the Twenty-First Century*, Westport, Conn.: Greenwood Publishing Group, 2003.

Morgan, Patrick M., *Deterrence: A Conceptual Analysis*, Beverly Hills, Calif.: Sage Publications, Inc., 1977.

———, *Deterrence: A Conceptual Analysis*, 2nd ed., Beverly Hills, Calif.: Sage Publications, Inc., 1983.

———, *Deterrence Now*, Cambridge: Cambridge University Press, 2003.

Mosey, Chris, "Nordic Enthusiasm for Nuclear-Free Zone Pleases USSR, but Not US," *Christian Science Monitor*, July 3, 1981. As of May 22, 2018:
http://www.csmonitor.com/1981/0703/070349.html

Mueller, Wolfgang, "The USSR and Permanent Neutrality in the Cold War," *Journal of Cold War Studies*, Vol. 18, No. 4, Fall 2016, pp. 148–179. As of May 22, 2018:
https://muse.jhu.edu/article/645911/pdf

Myers, Steven Lee, and Thom Shanker, "Bush Aides Say Russia Actions in Georgia Jeopardize Ties," *New York Times*, August 14, 2008.

"Nachal'nik General'nogo shtaba Vooruzhennyx Sil RF general armii Valerii Gerasimov provel telefonny razgovor s predsedatelem Voennogo Komiteta NATO generalom Petrom Pavelom" [Chief of the general staff of the armed forces Valerii Gerasimov conducted a telephone conversation with chairman of the NATO Military Committee Petr Pavel], March 3, 2017. As of May 24, 2018:
http://function.mil.ru/news_page/person/more.htm?id=12113548@egNews

Narang, Vipin, "What Does It Take to Deter? Regional Power Nuclear Postures and International Conflict," *Journal of Conflict Resolution*, Vol. 57, No. 3, June 2013, pp. 478–508.

Narinskii, Michail M., "The Soviet Union and the Berlin Crisis," in Francesca Gori and Silvio Pons, eds., *The Soviet Union and Europe in the Cold War, 1943–1953*, New York: St. Martin's Press, 1996.

National Security Council, *NSC 28: A Report to the National Security Council by the Executive Secretary on the Position of the United States with Respect to Scandinavia*, Washington, D.C.: National Security Council, August 26, 1948a.

———, *NSC 28/1: A Report to the President by the National Security Council on the Position of the United States with Respect to Scandinavia*, Washington, D.C.: National Security Council, September 3, 1948b.

———, *NSC 6006/1: U.S. Policy Toward Scandinavia (Denmark, Norway and Sweden)*, Washington, D.C.: National Security Council, April 6, 1960.

———, National Security Directive 26 (unclassified), October 2, 1989. As of May 24, 2018:
https://fas.org/irp/offdocs/nsd/nsd26.pdf

National Security Strategy of the United States of America, December 2017. As of May 24, 2018:
https://www.whitehouse.gov/wp-content/uploads/2017/12/NSS-Final-12-18-2017-0905.pdf

Natoli, Kristopher, "Weaponizing Nationality: An Analysis of Russia's Passport Policy in Georgia," *Boston University International Law Journal*, Vol. 28, Summer 2010, pp. 389–417.

Nilssen, Lawrence R., "Nordic NATO in Transition: Toward Turbulence in the 1990's?" *Airpower Journal*, Summer 1988. As of May 22, 2018:
http://www.airuniversity.af.mil/Portals/10/ASPJ/journals/Volume-02_Issue-1-4/1988_Vol2_No2.pdf

Nixon, Richard, "Address at the Commemorative Session of the North Atlantic Council," April 10, 1969, American Presidency Project. As of May 22, 2018:
http://www.presidency.ucsb.edu/ws/index.php?pid=1992&st=atlantic&st1=

"Nordic Nuclear Weapon–Free Zone," House of Lords Debate, Vol. 424 cc895-7, October 27, 1981. As of May 22, 2018:
http://hansard.millbanksystems.com/lords/1981/oct/27/nordic-nuclear-weapon-free-zone

North Atlantic Military Committee, *Memorandum for the Members of the Military Committee in Permanent Session: Study on Alert Measures in Support of Berlin Contingency Plans*, October 18, 1962. As of May 24, 2018:
https://www.nato.int/nato_static_fl2014/assets/pdf/pdf_archives/19621018-DP-MCM-120-62-ENG.pdf

———, *Final Decision on MC 14/3: A Report by the Military Committee to the Defence Planning Committee on Overall Strategic Concept for the Defense of the North Atlantic Treaty Organization Area*, January 16, 1968. As of May 22, 2018:
http://www.nato.int/docu/stratdoc/eng/a680116a.pdf

North Atlantic Treaty Organization, "Wales Summit Declaration," September 5, 2014. As of May 24, 2018:
https://www.nato.int/cps/ic/natohq/official_texts_112964.htm

———, "NATO Force Integration Units," fact sheet, September 2015. As of May 24, 2018:
http://www.nato.int/nato_static_fl2014/assets/pdf/pdf_2015_09/20150901_150901-factsheet-nfiu_en.pdf

———, "Press Conference by NATO Secretary General Jens Stoltenberg Following the North Atlantic Council Meeting at the Level of NATO Defence Ministers," June 14, 2016a. As of May 24, 2018:
https://www.nato.int/cps/su/natohq/opinions_132349.htm?selectedLocale=en

———, "Warsaw Summit Communiqué," July 9, 2016b. As of May 24, 2018:
https://www.nato.int/cps/en/natohq/official_texts_133169.htm

———, "Collective Defense—Article 5," March 22, 2017a. As of May 29, 2018:
http://www.nato.int/cps/en/natohq/topics_110496.htm

———, "Remarks by NATO Secretary General Jens Stoltenberg at the Elliott School of International Affairs, George Washington University," April 13, 2017b. As of May 24, 2018:
https://www.nato.int/cps/en/natohq/opinions_143137.htm

———, "Press Conference by NATO Secretary General Jens Stoltenberg Ahead of the Meeting of NATO Heads of State and Government," May 24, 2017c. As of May 24, 2018:
https://www.nato.int/cps/su/natohq/opinions_144081.htm?selectedLocale=en

———, "Boosting NATO's Presence in the East and Southeast," August 11, 2017d. As of May 24, 2018:
https://www.nato.int/cps/en/natohq/topics_136388.htm

"North Atlantic Treaty Organization," *Military Balance*, Vol. 61, No. 1, 1961, pp. 9–21. As of May 22, 2018:
http://www.tandfonline.com/doi/pdf/10.1080/04597226108459678

Obama, Barack, "Remarks of President Obama at 25th Anniversary of Freedom Day," June 4, 2014a. As of May 29, 2018:
https://obamawhitehouse.archives.gov/the-press-office/2014/06/04/remarks-president-obama-25th-anniversary-freedom-day

———, "Remarks at Nordea Concert Hall in Tallinn, Estonia," September 3, 2014b. As of May 24, 2018:
https://www.gpo.gov/fdsys/pkg/DCPD-201400640/pdf/DCPD-201400640.pdf

———, "Remarks by President Obama and Leaders of Baltic States in Multilateral Meeting," September 3, 2014c. As of May 24, 2018:
https://obamawhitehouse.archives.gov/the-press-office/2014/09/03/remarks-president-obama-and-leaders-baltic-states-multilateral-meeting

Oliker, Olga, *Russia's Nuclear Doctrine: What We Know, What We Don't, and What That Means*, Washington, D.C.: Center for Strategic and International Studies, May 2016. As of May 24, 2018:
https://csis-prod.s3.amazonaws.com/s3fs-public/publication/160504_Oliker_RussiasNuclearDoctrine_Web.pdf

Oliker, Olga, Christopher S. Chivvis, Keith Crane, Olesya Tkacheva, and Scott Boston, *Russian Foreign Policy in Historical and Current Context: A Reassessment*, Santa Monica, Calif.: RAND Corporation, PE-144-A, 2015. As of May 22, 2018:
https://www.rand.org/pubs/perspectives/PE144.html

Olsen, John Andreas, "Introduction: The Quest for Maritime Supremacy," in John Andreas Olsen, ed., *NATO and the North Atlantic: Revitalising Collective Defense*, Whitehall Paper No. 87, Abingdon, England: Routledge, 2016, pp. 3–7. As of May 22, 2018:
http://www.tandfonline.com/doi/pdf/10.1080/02681307.2016.1291017?needAccess=true

"Oral History: Tariq Aziz." *PBS Frontline*, 1995. As of May 24, 2018:
http://www.pbs.org/wgbh/pages/frontline/gulf/oral/aziz/1.html

"Oral History: Wafic Al Samarrai." *PBS Frontline*, 1995. As of May 24, 2018:
http://www.pbs.org/wgbh/pages/frontline/gulf/oral/samarrai/1.html

Ørvik, Nils, "Soviet Approaches on NATO's Northern Flank," *International Journal*, Vol. 20, No. 1, Winter 1964–1965, pp. 54–67. As of May 22, 2018:
https://www.jstor.org/stable/pdf/40199383.pdf

———, "Norway: Deterrence Versus Nonprovocation," in Nils Ørvik, ed., *Semialignment and Western Security*, New York: St. Martin's Press, 1986.

Osburg, Jan, Stephen J. Flanagan, and Marta Kepe, "How to Deter NATO's Greatest Fear: A Russian Invasion of the Baltic States," *National Interest*, November 22, 2016. As of May 24, 2018: http://nationalinterest.org/blog/the-buzz/if-you-want-peace-prepare-resistance"-18480

Oznobischev, Sergei, "Peretyagivanie Mira—chast' 1, Obschie Vyzovy I Ugrozy Vazhnee Krizisa na Ukraine" [Tug of peace—Part 1, common challenges and threats are more important than the Ukrainian crisis], *Voenno-Promyshlenny Kur'er*, July 23, 2014. As of May 24, 2018: http://www.vpk-news.ru/articles/21127

———, "Russia and NATO: From the Ukrainian Crisis to the Renewed Interaction," in Alexei Arbatov and Sergei Oznobishchev, eds., *Russia: Arms Control, Disarmament and International Security*, Moscow: IMEMO, 2016.

Pape, Robert A., Bombing *to Win: Air Power and Coercion in War*, Ithaca, N.Y.: Cornell University Press, 2014.

Parrish, Thomas, *Berlin in the Balance 1945–1949: The Blockade, the Airlift, the First Major Battle of the Cold War*, Reading, Mass.: Addison-Wesley, 1998.

Paul, T. V., "Complex Deterrence: An Introduction," in T. V. Paul, Patrick M. Morgan, and James J. Wirtz, eds., *Complex Deterrence: Strategy in the Global Age*, Chicago: University of Chicago Press, 2009.

Person, Robert, "6 Reasons Not to Worry About Russia Invading the Baltics," *Washington Post*, November 12, 2015. As of May 24, 2018: https://www.washingtonpost.com/news/monkey-cage/wp/2015/11/12/6-reasons-not-to-worry-about-russia-invading-the-baltics/?utm_term=.43eab43c08cd

"Peskov rasskazal pro ozhidaniia ot administratsii Trampa" [Peskov addressed expectations of the Trump administration], *Rossiiskaya Gazeta*, December 21, 2016.

"Peskov: RF bespokoit rasshirenie al'iansov k ee granitsam, a ne otnosheniia sosedei s SSHA" [Peskov: Russia is concerned about the enlargement of alliances to its borders, but not about relations of its neighbors with the U.S.], August 1, 2017. As of May 24, 2018: http://tass.ru/politika/4453642

Petersen, Nikolaj, "SAC at Thule: Greenland in the U.S. Polar Strategy," *Journal of Cold War Studies*, Vol. 13, No. 2, Spring 2011, pp. 90–115. As of May 22, 2018: http://www.mitpressjournals.org/doi/pdf/10.1162/JCWS_a_00138

Post, Jerrold M., "The Defining Moment of Saddam's Life: A Political Psychology Perspective on the Leadership and Decision Making of Saddam Hussein During the Gulf Crisis," in Stanley A. Renshon, ed., *The Political Psychology of the Gulf War: Leaders, Publics, and the Process of Conflict*, Pittsburgh: University of Pittsburgh Press, 1993.

Press, Daryl G., "The Credibility of Power: Assessing Threats During the 'Appeasement' Crises of the 1930s," *International Security*, Vol. 29, No. 3, Winter 2004–2005, pp. 136–139, 168–169.

———, *Calculating Credibility: How Leaders Assess Military Threats*, Ithaca, N.Y.: Cornell University Press, 2007.

Quackenbush, Stephen L., "General Deterrence and International Conflict: Testing Perfect Deterrence Theory," *International Interactions*, Vol. 36, No. 1, January 2010, pp. 60–85.

———, "Deterrence Theory: Where Do We Stand?" *Review of International Studies*, No. 31, 2011, pp. 741–762.

Quester, George H., *Deterrence Before Hiroshima: The Airpower Background of Modern Strategy*, New York: John Wiley and Sons, 1966.

Radin, Andrew, and Clinton Bruce Reach, *Russian Views of the International Order*, Santa Monica, Calif.: RAND Corporation, RR-1826-OSD, 2017. As of May 24, 2018: https://www.rand.org/pubs/research_reports/RR1826.html

Rauchhaus, Robert, "Evaluating the Nuclear Peace Hypothesis: A Quantitative Approach," *Journal of Conflict Resolution*, Vol. 53, No. 2, April 2009, pp. 258–277.

Reagan, Ronald, "Remarks on the 35th Anniversary of the North Atlantic Alliance," May 31, 1984, American Presidency Project. As of May 22, 2018: http://www.presidency.ucsb.edu/ws/index.php?pid=39982&st=atlantic&st1=

———, "Written Responses to Questions Submitted by the Swedish Newspaper Svenska Dagbladet," September 22, 1987, American Presidency Project. As of May 22, 2018: http://www.presidency.ucsb.edu/ws/index.php?pid=33461&st=scandinavia&st1=

Reed, Anderson R., Patrick J. Ellis, Antonio M. Paz, Kyle A. Reed, Lendy Reenegar, and John T. Vaughn, *Strategic Landpower and a Resurgent Russia: An Operational Approach to Deterrence*, Carlisle, Penn.: U.S. Army War College Strategic Studies Institute, 2016.

Reuters staff, "Merkel Pledges NATO Will Defend Baltic Member States," Reuters, August 18, 2014. As of May 24, 2018: https://www.reuters.com/article/uk-ukraine-crisis-baltics-merkel/merkel-pledges-nato-will-defend-baltic-member-states-idUKKBN0GI1J420140818

Rice, Condoleezza, "Remarks with Georgian President Mikheil Saakashvili," July 10, 2008, Department of State Archives. As of May 25, 2018: https://2001-2009.state.gov/secretary/rm/2008/07/106912.htm

———, *No Higher Honor: A Memoir of My Years in Washington*, New York: Crown Publishers, 2011.

Rid, Thomas, "Deterrence Beyond the State: The Israeli Experience," *Contemporary Security Policy*, Vol. 33, No. 1, 2012, pp. 124–147.

Riste, Olav, "NATO's Northern Frontline in the 1980s," in Olav Njøstad, ed., *The Last Decade of the Cold War: From Conflict Escalation to Conflict Transformation*, Abingdon, England: Taylor and Francis, 2005, pp. 301–310.

"Rossiyane stali khuzhe otnosit'sya k Trampu" [Russians' opinion of Trump worsened], *Rossiiskaya Gazeta*, August 4, 2017.

Ruiz Palmer, Diego A., "The NATO-Warsaw Pact Competition in the 1970s and 1980s: A Revolution in Military Affairs in the Making or the End of a Strategic Age?" *Cold War History*, Vol. 14, No. 4, 2014, pp. 533–573. As of May 22, 2018: http://www.nato.int/nato_static_fl2014/assets/pdf/pdf_history/20161212_E2-Ruiz-Palmer-2014-NATO-Warsaw-Pact-competition.pdf

Russett, Bruce M., "The Calculus of Deterrence," *Journal of Conflict Resolution*, Vol. 7, No. 2, June 1963, pp. 97–109.

"Russia Accused of Estonia Airspace Violations as Finland Signs Defense Pact with US," *Deutsche Welle*, October 8, 2016. As of May 24, 2018: http://www.dw.com/en/russia-accused-of-estonia-airspace-violations-as-finland-signs-defense-pact-with-us/a-35994791

Sagan, Scott D. "Two Renaissances in Nuclear Security Studies," *H-Diplo/ISSF Forum*, No. 2, 2014, pp. 2–10.

Saideman, Stephen M., and Marie-Joëlle Zahar, *Intra-State Conflict, Governments and Security: Dilemmas of Deterrence and Assurance*, Abingdon, England: Routledge, 2008.

Schelling, Thomas C., *Arms and Influence*, New Haven, Conn.: Yale University Press, 1966.

———, *The Strategy of Conflict*, Cambridge, Mass.: Harvard University Press, 1980.

Schneider, Barry R. *Deterrence and Saddam Hussein: Lessons from the 1990–1991 Gulf War*, Maxwell Air Force Base, Ala.: US Air Force Counterproliferation Center, 2009.

Schneider, Barry R., and Patrick D. Ellis, eds., *Tailored Deterrence: Influencing States and Groups of Concern*, Maxwell Air Force Base, Ala.: U.S. Air Force Counterproliferation Center, 2011. As of May 24, 2018:
http://www.au.af.mil/au/cpc/assets/tailor_deterence.pdf

Schneider, Mark B., "Escalate to De-Escalate," *Proceedings*, February 2017.
https://www.usni.org/magazines/proceedings/2017-02/escalate-de-escalate

Schöpflin, George, "NATO and the Nordic Balance," *World Today*, Vol. 22, No. 3, March 1966. As of May 22, 2018:
https://www.jstor.org/stable/pdf/40393837.pdf?refreqid=excelsior:aa895573f392968aa165f0b3adf34011

Schroeder, Paul W., *Systems, Stability, and Statecraft: Essays on the International History of Modern Europe*, New York: Palgrave Macmillan, 2004.

Sciolino, Elaine, and Michael R. Gordon, "U.S. Gave Iraq Little Reason Not to Mount Kuwait Assault." *New York Times*, September 23, 1990. As of May 24, 2018:
http://www.nytimes.com/1990/09/23/world/confrontation-in-the-gulf-us-gave-iraq-little-reason-not-to-mount-kuwait-assault.html?pagewanted=all

Sechser, Todd S., and Matthew Furhmann, "Crisis Bargaining and Nuclear Blackmail," *International Organization*, Vol. 67, Winter 2013, pp. 173–95.

———, "Signaling Alliance Commitments: Hand-Tying and Sunk Costs in Extended Nuclear Deterrence," *American Journal of Political Science*, Vol. 58, No. 4, October 2014, pp. 919–935.

Shimshoni, Jonathan, *Israel and Conventional Deterrence: Border Warfare from 1953 to 1970*, Ithaca, N.Y.: Cornell University Press, 1988.

Shlaim, Avi. *The United States and the Berlin Blockade, 1948–1949: A Study in Crisis Decision-Making*, Berkeley: University of California Press, 1983.

Shlapak, David A., and Michael W. Johnson, *Reinforcing Deterrence on NATO's Eastern Flank: Wargaming the Defense of the Baltics*, Santa Monica, Calif.: RAND Corporation, RR-1253-A, 2016. As of May 24, 2018:
https://www.rand.org/pubs/research_reports/RR1253.html

Signorino, Curtis S., and Ahmer Tarar, "A Unified Theory and Test of Extended Immediate Deterrence," *American Journal of Political Science*, Vol. 50, No. 3, July 2006, pp. 586–605.

Simmons, Katie, Bruce Stokes, and Jacob Poushter, "NATO Publics Blame Russia for Ukrainian Crisis, but Reluctant to Provide Military Aid," Pew Research Center, June 10, 2015. As of May 24, 2018:
http://assets.pewresearch.org/wp-content/uploads/sites/2/2015/06/Pew-Research-Center-Russia-Ukraine-Report-FINAL-June-10-2015.pdf

Smyser, W. R., *Kennedy and the Berlin Wall*, Lanham, Md.: Rowman and Littlefield Publishers, 2009.

Snyder, Glenn H., *Deterrence by Denial and Punishment*. Princeton, N.J.: Center of International Studies, January 1959.

Snyder, Jack, and Erica D. Borghard, "The Cost of Empty Threats: A Penny, Not a Pound," *American Political Science Review*, Vol. 105, No. 3, August 2011, pp. 437–456.

Soderlind, Rolf, "Norwegian Airfield Crucial in World War III Scenario," UPI, April 11, 1985. As of May 22, 2018:
http://www.upi.com/Archives/1985/04/11/Norwegian-airfield-crucial-in-World-War-III-scenario/3379482043600/

Sohlberg, Ragnhild, *Analysis of Ground Force Structures on NATO's Northern Flank*, Santa Monica, Calif.: RAND Corporation, N-1315-MRAL, 1980. As of May 24, 2018:
https://www.rand.org/pubs/notes/N1315.html

"The Soviet Union," *Military Balance*, Vol. 90, No. 1, 1990, pp. 28–43. As of May 22, 2018:
http://www.tandfonline.com/doi/pdf/10.1080/04597229008460018

"Spiker Kongressa: SSHA podderzhat Estoniyu v otnoshenii ugroz s vostoka" [Speaker of Congress: USA will support Estonia with regard to threats from the east], April 22, 2017. As of May 24, 2018:
http://tass.ru/mezhdunarodnaya-panorama/4204056

"St. Petersburg International Economic Forum Plenary Meeting," June 2, 2017. As of August 18, 2017:
http://en.kremlin.ru/events/president/news/54667

Steavenson, Wendell, "Marching Through Georgia: Has Mikheil Saakashvili Overreached?" *New Yorker*, December 15, 2008. As of June 8, 2018:
https://www.newyorker.com/magazine/2008/12/15/marching-through-georgia

Stein, Janice Gross, "Threat-Based Strategies of Conflict Management: Why Did They Fail in the Gulf?" in Stanley A. Renshon, ed., *The Political Psychology of the Gulf War: Leaders, Publics, and the Process of Conflict*, Pittsburgh: University of Pittsburgh Press, 1993, pp. 121–153.

———, "Rational Deterrence Against 'Irrational' Adversaries?" in T. V. Paul, Patrick M. Morgan, and James J. Wirtz, eds., *Complex Deterrence: Strategy in the Global Age*, Chicago: University of Chicago Press, 2009.

Stein, Janice Gross, and David A. Welch, "Rational and Psychological Approaches to the Study of International Conflict: Comparative Strengths and Weaknesses," in Nehemia Geva and Alex Mintz, eds., *Decisionmaking on War and Peace*, Boulder, Colo.: Lynne Rienner, 1997.

Steiner, André, "From Soviet Occupation Zone to 'New Eastern State,'" in Hartmut Berghoff and Uta Andrea Balbier, eds., *The East German Economy, 1945–2010: Falling Behind or Catching Up?* Cambridge: Cambridge University Press, 2013.

Stent, Angela, *The Limits of Partnership: U.S.-Russian Relations in the Twenty-First Century*, Princeton, N.J.: Princeton University Press, 2014.

Stern, David L., "Security Group Refuses to Back Russia's Actions," *New York Times*, August 20, 2008.

Steury, Donald P., ed., *On the Front Lines of the Cold War: Documents on the Intelligence War in Berlin, 1946 to 1961*, Washington, D.C.: Center for the Study of Intelligence, 1999.

Stone, J. "Conventional Deterrence and the Challenge of Credibility," *Contemporary Security Policy*, Vol. 33, No. 1, 2010, pp. 108–123.

Sullivan, William K., *Soviet Strategy and NATO's Northern Flank*, Carlisle, Penn.: Strategic Studies Institute, U.S. Army War College, May 1, 1978. As of May 22, 2018:
http://www.dtic.mil/dtic/tr/fulltext/u2/a054369.pdf

"Swedes Mock Soviet on Atom-Free Zone," *New York Times*, November 8, 1981. As of May 22, 2018:
http://www.nytimes.com/1981/11/08/world/swedes-mock-soviet-on-atom-free-zone.html

Sytas, Andrius, "NATO War Game Defends Baltic Weak Spot for First Time," Reuters, June 18, 2017. As of May 24, 2018:
https://www.reuters.com/article/us-nato-russia-suwalki-gap/nato-war-game-defends-baltic-weak-spot-for-first-time-idUSKBN1990L2

Taubman, William, *Khrushchev: The Man and His Era*, New York: W. W. Norton and Company, 2003.

Teal, John J., Jr., "Europe's Northernmost Frontier," *Foreign Affairs*, January 1951, pp. 263–275. As of May 22, 2018:
https://www.foreignaffairs.com/articles/northern-europe/1951-01-01/europes-northernmost-frontier

Tebin, Prokhor, *A Tranquilizer with a Scent of Gunpowder: The Balance Between Russian and NATO Forces in Eastern Europe After 2014*, Valdai Papers No. 70, July 2017. As of May 24, 2018:
http://valdaiclub.com/files/14970/

"Tensions Rise on Norwegian-Soviet Border," *New York Times*, December 7, 1986. As of May 22, 2018:
http://www.nytimes.com/1986/12/07/world/tensions-rise-on-norwegians-soviet-border.html

Terrill, W. Andrew, *Escalation and Intrawar Deterrence During Limited Wars in the Middle East*, Carlisle, Penn.: Strategic Studies Institute, U.S. Army War College, 2009.

Tilghman, Andrew, and Oriana Pawlyk, "U.S. vs. Russia: What a War Would Look Like Between the World's Most Fearsome Militaries," *Military Times*, October 5, 2015.

Trachtenberg, Marc, *A Constructed Peace: The Making of the European Settlement, 1945–1963*, Princeton, N.J.: Princeton University Press, 1999.

"Transcript: Donald Trump on NATO, Turkey's Coup Attempt and the World," *New York Times*, July 21, 2016.

Truman, Harry S., "Letter to Committee Chairmen on the Need for Continuing Aid to Denmark," July 25, 1952, American Presidency Project. As of May 22, 2018:
http://www.presidency.ucsb.edu/ws/index.php?pid=14214&st=denmark&st1=

Trump, Donald J., "Remarks by President Trump at NATO Unveiling of the Article 5 and Berlin Wall Memorials—Brussels, Belgium," May 25, 2017.

Tunander, Ola, "The Logic of Deterrence," *Journal of Peace Research*, Vol. 26, No. 4, November 1989, pp. 353–365. As of May 22, 2018:
http://www.jstor.org/stable/pdf/423658.pdf

"The United States and the Soviet Union," *Military Balance*, Vol. 81, No. 1, 1981, pp. 3–15. As of May 22, 2018:
http://www.tandfonline.com/doi/pdf/10.1080/04597228108459912

U.S. Army, "Countdown to 75: US Army Europe and REFORGER." March 22, 2017. As of May 25, 2018:
https://www.army.mil/article/184698/countdown_to_75_us_army_europe_and_reforger

U.S. Army Europe, "Exercise Saber Strike 14 Demonstrates International Cooperation," June 10, 2014. As of May 24, 2018:
https://www.army.mil/article/127664/exercise_saber_strike_14_demonstrates_international_cooperation

U.S. Army Europe Public Affairs, "Exercise Saber Strike 2012 Demonstrates International Cooperation in Action," March 15, 2012. As of May 24, 2018:
https://media.defense.gov/2018/May/03/2001911959/-1/-1/0/05152012%20EXERCISE%20SABER%20STRIKE%202012%20DEMONSTRATES%20INTERNATIONAL%20COOPERATION%20IN%20ACTION.PDF

———, "Exercise Saber Strike 15 Demonstrates International Cooperation Capabilities," June 1, 2015. As of May 24, 2018:
https://www.army.mil/article/149596/exercise_saber_strike_15_demonstrates_international_cooperation_capabilities

U.S. Department of Defense, Office of the Under Secretary of Defense (Comptroller), *European Reassurance Initiative: Department of Defense Budget, Fiscal Year (FY) 2016*, Washington, D.C.: U.S. Department of Defense, February 2015.

———, *European Reassurance Initiative: Department of Defense Budget, Fiscal Year (FY) 2018*, Washington, D.C.: U.S. Department of Defense, May 2017.

U.S. Department of State, U.S. Embassies and Consulates in Germany, "Visits of U.S. Presidents to Germany Since 1945," n.d. As of April 21, 2017:
https://de.usembassy.gov/our-relationship/policy-history/official-visits/

———, *Defense of Greenland*, April 9, 1941, Library of Congress. As of May 22, 2018:
https://www.loc.gov/law/help/us-treaties/bevans/b-dk-ust000007-0107.pdf

———, "Georgia Train and Equip Program (GTEP)," February 1, 2003. As of May 25, 2018:
https://2001-2009.state.gov/r/pa/ei/pix/b/eur/18737.htm

U.S. Embassy in Estonia, "U.S.-Estonia Relations," n.d. As of May 24, 2018:
https://ee.usembassy.gov/our-relationship/policy-history/us-estonia-relations/

U.S. European Command, "Navy Ship Encounters Aggressive Russian Aircraft in Baltic Sea," U.S. Department of Defense, April 13, 2016a. As of May 29, 2018:
https://www.defense.gov/News/Article/Article/720536/navy-ship-encounters-aggressive-russian-aircraft-in-baltic-sea/

U.S. Marines, Combat Logistics Regiment 2, "U.S. NATO Allies Conduct Large-Scale Exercise to Defend Baltics," June 20, 2016. As of May 24, 2018:
https://www.marines.mil/News/News-Display/Article/804914/us-nato-allies-conduct-large-scale-exercise-to-defend-baltics/

Vandiver, John, "Breedlove: NATO Must Redefine Responses to Unconventional Threats," *Stars and Stripes*, July 31, 2014.

Van Elsuwege, Peter, *Former Soviet Republics to EU Member States: A Legal and Political Assessment of the Baltic States' Accession to the EU*: Leiden: Brill, 2008.

Vartanyan, Olesya, and Ellen Barry, "Ex-Diplomat Says Georgia Started War Against Russia," *New York Times*, November 25, 2008.

Ven Bruusgaard, Kristin, "Crimea and Russia's Strategic Overhaul," *Parameters*, Fall 2014, pp. 81–90.

———, "Russian Strategic Deterrence," *Survival*, Vol. 58, No. 4, November–December 2014, pp. 7–26.

———, "The Myth of Russia's Lowered Nuclear Threshold," *War on the Rocks*, September 22, 2017. As of May 24, 2018:
https://warontherocks.com/2017/09/the-myth-of-russias-lowered-nuclear-threshold/

Vendil Pallin, Carolina, and Fredrik Westerlund, "Russia's War in Georgia: Lessons and Consequences," *Small Wars and Insurgencies*, Vol. 20, No. 2, June 2009, pp. 400–424.

Vice, Margaret, "Russians Remain Confident in Putin's Global Leadership," Pew Research Center, June 20, 2017. As of May 24, 2018:
http://www.pewglobal.org/2017/06/20/president-putin-russian-perspective/

Vinocur, John, "Brandt's Soviet Visit Troubles Schmidt and NATO," *New York Times*, July 13, 1981. As of May 22, 2018:
http://www.nytimes.com/1981/07/13/world/brandt-s-soviet-visit-troubles-schmidt-and-nato.html

"Vladimir Putin Meets with Members of the Valdai Discussion Club. Transcript of the Plenary Session of the 13th Annual Meeting," October 27, 2016. As of May 24, 2018:
http://valdaiclub.com/events/posts/articles/vladimir-putin-took-part-in-the-valdai-discussion-club-s-plenary-session/

"Vystupleniie Nachal'nika Genshtaba VS RF Generala Armii Valeriya Gerasimova na Konferentsii MCIS-2016" [Presentation of the chief of the general staff of the armed forces, Valery Gerasimov, at the MCIS-2016 Conference]," April 26, 2017. As of May 24, 2018:
http://mil.ru/mcis/news/more.htm?id=12120704@cmsArticle

Weede, Erich, "Extended Deterrence by Superpower Alliance," *Journal of Conflict Resolution*, Vol. 27, No. 2, June 1983, pp. 231–254.

Weisiger, Alex, and Keren Yarhi-Milo, "Revisiting Reputation: How Past Actions Matter in International Politics," *International Organization*, Vol. 69, No. 2, Spring 2015, pp. 473–495.

Weiss, Andrew S., "Are Russian Protests a Threat to Putin?" KCRW Radio, June 12, 2017. As of May 24, 2018:
http://carnegieendowment.org/2017/06/12/are-russian-protests-threat-to-putin-pub-71231

The White House, Office of the Press Secretary, "Fact Sheet: European Reassurance Initiative and Other U.S. Efforts in Support of NATO Allies and Partners," June 3, 2014. As of May 24, 2018:
https://obamawhitehouse.archives.gov/the-press-office/2014/06/03/fact-sheet-european-reassurance-initiative-and-other-us-efforts-support-

———, "Remarks by President Trump and President Iohannis of Romania in a Joint Press Conference," June 9, 2017. As of May 24, 2018:
https://www.whitehouse.gov/the-press-office/2017/06/09/remarks-president-trump-and-president-iohannis-romania-joint-press

Williamson, Richard D., *First Steps Toward Détente: American Diplomacy in the Berlin Crisis, 1958–1963*, Lanham, Md.: Lexington Books, 2012.

Wolf, Barry, *When the Weak Attack the Strong: Failures of Deterrence*, Santa Monica, Calif.: RAND Corporation, N-3261-A, 1991. As of December 15, 2017:
https://www.rand.org/pubs/notes/N3261.html

Woods, Kevin M., David D. Pakki, and Mark E. Stout, eds., *The Saddam Tapes: The Inner Workings of a Tyrant's Regime, 1978–2001*, Cambridge: Cambridge University Press, 2011.

Woods, Kevin M., and Mark E. Stout, "Saddam's Perceptions and Misperceptions: The Case of 'Desert Storm,'" *Journal of Strategic Studies*, Vol. 33, No. 1, February 2010, pp. 5–41.

Work, Bob, "The Third U.S. Offset Strategy and Its Implications for Partners and Allies," U.S. Department of Defense, January 28, 2015. As of May 24, 2018:
https://www.defense.gov/News/Speeches/Speech-View/Article/606641/the-third-us-offset-strategy-and-its-implications-for-partners-and-allies/

Zagare, Frank C., and D. Marc Kilgour, "Asymmetric Deterrence," *International Studies Quarterly*, Vol. 37, No. 1, March 1993, pp. 1–27. As of May 22, 2018:
http://www.jstor.org/stable/pdf/2600829.pdf

———, "Deterrence Theory and the Spiral Model Revisited," *Journal of Theoretical Politics*, Vol. 10, No. 1, January 1998, pp. 59–87.

———, *Perfect Deterrence*, Cambridge: Cambridge University Press, 2000.

"Zasedanie Kollegii Federal'noi Sluzhby Bezobasnosti" [Meeting of the Collegium of the Federal Security Service], February 16, 2017. As of May 24, 2018:
http://kremlin.ru/events/president/news/page/39

"Zayavlenie MID Rossii v Svyazi s Yubileinymi Datami v Otnosheniiakh Rossia-NATO" [Statement of the Ministry of Foreign Affairs in commemoration of key dates in Russia-NATO relations], April 26, 2017. As of May 24, 2018:
http://www.mid.ru/ru/press_service/spokesman/official_statement/-/asset_publisher/t2GCdmD8RNIr/content/id/2767662

Lightning Source UK Ltd.
Milton Keynes UK
UKHW050310230721
387619UK00002B/27